A. A. Berlin
St. A. Volfson
N. S. Enikolopian
S. S. Negmatov

Principles of Polymer Composites

Principles of Polymer Composites

by

Prof. Dr. Alexander A. Berlin
Dr. Stanislav A. Volfson
Prof. Dr. Nicolai S. Enikolopian
Moscow

and

Prof. Dr. S. S. Negmatov
Taschkent

With 40 Figures

Akademie-Verlag Berlin
1986

Die Originalausgabe erscheint im Springer-Verlag
Berlin · Heidelberg · New York · Tokyo

Vertrieb ausschließlich für die DDR und die sozialistischen Länder

ISBN 3-540-15051-X Springer-Verlag Berlin · Heidelberg · New York · Tokyo
ISBN 0-387-15051-X Springer-Verlag New York · Heidelberg · Berlin · Tokyo

ISBN 3-05-500136-2

Erschienen im Akademie-Verlag Berlin, DDR-1086 Berlin, Leipziger Straße 3—4
Lizenznummer: 202 · 100/526/86
Printed in the German Democratic Republic
Gesamtherstellung: VEB Druckerei „Thomas Müntzer", 5820 Bad Langensalza
Umschlaggestaltung: Karl Salzbrunn
LSV 1275
Bestellnummer: 763 482 8 (6905)
14800

Preface

Polymer composites represent a field of intense and growing interest to consumers and producers of plastics. They are used to solve the most acute problems: energy and oil conservation, improvement of the properties of polymer materials, and the increase of their use. Even special problems, such as inflammability of plastics and industrial wastes, are closely connected with polymer composites.

The achievements in this field are well known. Polymer composites have been widely used in building, furniture, electric appliances, cars, and other fields. Their production is growing at a higher rate than that of polymers as a whole.

In the present book the emphasis is put on the principles that may become the foundation of designing new highly effective composites. The authors analyze their favorable properties as compared to "unfilled" polymers, as well as the means to improve moldability and strength. Economical and technical problems are examined.

Special attention is paid to the matching of the components, development of technological processes of composites production, and to new ideas in the field.

Fundamental and practical aspects of calculating properties and structure of composites are examined.

The scope of the book does not include composites based on continuous reinforcing fibers, polymeric concretes, nor other special-purpose materials in which polymers are used to modify the properties of inorganic materials.

The book reflects mainly ideas developed at the Institute of Chemical Physics of the USSR Academy of Sciences, but it also contains a review of the latest works in the field.

Moscow, September 1985

A. A. Berlin, S. A. Volfson,
N. S. Enikolopian, S. S. Negmatov

Contents

List of Symbols

Definition of Indices

c: composite
f: fiber
fl: dispersed filler
m: matrix

Definition of Terms

A: work of rupture
E: Young's modulus
F: distribution function
G: shear modulus
I: mixture index
K: bulk modulus
$K^{\pm}$: coefficient of concentration of stress (tension-compression)
K^{τ}: coefficient of concentration of stress (shear)
L: gage length
M: moment of volume distribution function
N: number of filler particles
P: loading
P(x): probability
Q(l) fixed function of distribution of fiber length
S: pliability
a: thickness of glass band; impact strength
b: width of glass band
c: effective crack size
d: diameter
g: parameter of band shrinkage
h: amount of material in neck
l: fiber length
l_c: critical fiber length
q: relative part of dispersive phase
p: pore dimension
r: radius
r_0: distance between particles
α: characteristic relation of diameter to thickness

List of Symbols

α, β: parameters of distribution
γ: effective surface energy of fracture
δ: noneffective fiber length
ε: relative elongation at rupture; deformation; strain
ε^*: ultimate elongation
ε_r: real elongation
η: viscosity
$\varkappa$: coefficient of electroconductivity
μ: Poisson ratio
ν: volume (weight) fraction
ν^*: maximum volume (weight) fraction
σ: stress
σ^*: ultimate stress
σ_{pl}: yield strength
σ_{sh}: shear stress
τ: shear strength
φ: angle of fiber bending
χ: permeability coefficient

Introduction

"Fillers are generally added to polymers to improve their physico-mechanical properties, processability and/or to lower the cost". (Common opinion)

It is long since this phrase began circulating in scientific papers, reviews and monographs, but it is not less urgent today and probably will not be less urgent tomorrow. Should we trust historical records, the first composite materials have been mentioned in . . . the Bible, namely: clay bricks reinforced by straw. Many researchers and engineers engaged in the polymer industry believe that the future belongs to composites, and that unfilled polymers will, by the end of this century, be looked upon as half-finished products, in the same way as rubber or carbamide resins are looked upon to-day.

Most thermosetting plastics are already used as high-filled compositions. Strong, high-modulus composites filled with reinforcing fibers have become a class of materials in itself. In many cases polymer coatings and adhesives are filled systems. On an ever greater scale, engineering thermoplastics are being filled with glass fibers for strength.

What territory remains uninvaded by fillers? Common thermoplastics? Fibers? Films? But PVC has long and successfully been filled with calcium carbonate in the manufacture of linoleum, decorative tiles, cable insulation. Filled polypropylene is becoming the most popular engineering plastic. In the technical literature there are reports of filled fibers and films.

Yet problems in filling do exist. Improvements of the whole complex of properties — physical, mechanical, rheological, physico-chemical, thermal, electrical, magnetic and others — are problems which can never be solved to ultimate satisfaction.

As regards common materials, the primary problem is to improve their economics. This may be solved either by lowering the cost of the material, or by improving processability and performance. Oil shortage and soaring energy costs are also factors stimulating the development of filled materials.

A comprehensive account of all mineral, organic and synthetic fillers currently in use has been published by Katz and Milewsky [1]. If to the impressive list of natural and synthetic materials reported as suitable fillers we add air, gases and certain liquids, we will be led to the conclusion that almost all materials, both natural ad man-made, may in one form or another be blended with a polymer matrix to result in a useful composite. Our modest contribution to this list includes plastic and rubber wastes as very promising fillers, especially for common thermoplastics [2].

All fillers must be imparted the requisite shape and particle size distribution; moreover they must be appropriately distributed in the polymer matrix. In the case of hard fillers, their surfaces must be treated to remove moisture and adsorbed gases or to provide special transition layers. The modern technology of composites is based on two major techniques, viz. impregnation of the filler with a molten polymer and compounding of the filler with molten polymer. But this does not mean, of course, that new processes may not emerge. One recent example is filling in the course of the polymerization.

The current state of the art of polymer matrix composites does not allow one to discuss all the problems of interest from a unified point of view. The amount of data collected by now is too large to enable a comprehensive analysis of all aspects involved, within the scope of one monograph.

It is easy to postulate the necessity of mass-scale filling of thermoplastics on the grounds of the apparent tendencies for rising oil and energy prices. But anyone who accepts this idea must be prepared to answer at least three questions:

i — Are there ways to preserve, in the filled material, the plasticity and impact resistance inherent in unfilled thermoplastics?

ii — How to assure the same level of processability in the filled material as in the unfilled?

iii — How to blend the filler with the polymer matrix without extra power requirement and deterioration of the properties of the individual components?

These questions have gained acuteness over the last decade because of the increasing competition between polymeric and conventional structural materials. They have been repeatedly discussed in literature [3–5].

Still we are far from the aim of large-scale use of filled thermoplastics since their total output has not yet attained even 1% of the annual production of common thermoplastics. Many specialists hold that mineral fillers will improve rigidity and dimensional stability, but others insist that they will increase brittleness and melt viscosity. Fibrous fillers are, in addition, capable of improving strength and heat resistance but, apart from the above mentioned drawbacks, they make the material costlier.

There are reasons to assume that most plastics manufacturing companies have been carrying out intensive research in the field of composites during the past 15 years. In most cases the research has been based on the following three postulates:

i — A polymer matrix is any commercial polymer or copolymer;

ii — the basic composite production technique is *blending* of the filler with molten polymer in an appropriate mechanical device;

iii — since production and property control of plastics and composites are branches of science and technology by themselves, all a composite specialist has to do is to combine the available components and choose additives capable of modifying the properties of the interface layers.

What possibilities still remain unused? We will try to set forth the scientific and technical problems involved.

Theory

1. The principles of polymer reinforcement with dispersed fillers at low and high strain are not yet quite clear. While the addition of ultrafine-dispersed carbon black and silica to elastomers results in an almost all-round improvement of properties, it has so far been impossible to obtain a similar effect in plastics.
2. Calculations of the mechanical properties of polymer matrix composites in terms of linear-elastic mechanics are in sufficiently good agreement with experiment.

However, considerabl theoretical difficulties arise when medium and large filling ratios are considered, one of the main problems being the interaction of filler particles with each other.

3. In calculations of the ultimate strength characteristics of polymer composites one must consider the mechanisms of their failure which is a very complex phenomenon depending on a large number of factors. Their classification, arrangement according to their relative contributions and inter-relations are still on a very low level of development.
4. The impact resistance of polymer matrices filled with rigid particles or fibers is of particular interest to technologists, since it is, in fact, the weak point of most composite materials. The toughening of plastics by their modification with elastomers would seem to open up new ways towards solving this problem.
5. The papers reviewing the rheology of filled thermoplastics published so far have done little more than describe empirical relationships.
6. The interfacial layers in polymer matrix composites have long been attracting keen attention. Although literally all researchers in the field agree that these layers exhibit an extremely important influence on the rheological and mechanical properties as well as on the processability of the whole system, our knowledge of their organization is surprisingly meagre. The theoretical treatments usually content themselves with good and bad adhesion models. Experimental studies have been focused mainly on the effects of silane coupling agents.

Technology

1. As generally in the history of technology, the advances in this field are far ahead of the theoretical treatments of polymer composites. Glass-fiber-reinforced plastics are already finding application in the manufacture of critical engineering components. The technologists' efforts have been largely channeled into the optimization of known systems.
2. In reinforced plastics the problem reduces to preserving the initial aspect ratio of the reinforcement in finished parts, assuring the optimum orientation, and improving the adhesion between reinforcement and matrix.
3. The possibilities of blending and compounding technologies for obtaining a finely and adequately organized structure of polymer composites are severely limited. If we refer to the history of rubber-modified polystyrene for analogy, we will conclude that the use of the "filling-during-polymerization" technique opens up broad possibilities in this respect.
4. The diversity of existing methods of chemical grafting of macrochains to the filler surface is in contradiction with the scarce published data about the properties of the resultant composites.
5. The versatility of silane coupling agents in filled elastomers, thermosets and thermoplastics challenges the technologist to seek for alternative agents that are cheaper or more effective.
6. The potential for designing polymer matrices for obtaining composites with a predetermined set of properties is still in its virginal state. There are considerable resources in this respect, both as regards the control of composite properties and improvement of their rheological parameters.

While this book bears the broad title of PRINCIPLES OF POLYMER COMPOSITES, we have decided to focus our attention mainly on the still unsolved problems in the theory and practice of polymer composites.

Therefore, in Chapter I we analyze the current theoretical conceptions of reinforcement of polymer matrices by fillers. Emphasis is laid on impact strength, as well as on possible ways of elasticization of the filler.

Chapter II reviews the current status of the theory of calculation of mechanical properties of different polymer composites, including examples which are easy or difficult to calculate. Attention is centered on fiber-filled composites, the theory of which is in a more advanced state. Examples of the calculation of certain physical and thermophysical properties of polymer composites are given.

Chapter III is a critical review of the principles underlying the current technologies of polymer composite production. Among the surface treatment methods, emphasis is laid on the chemical grafting of macromolecules to the surface (polymerization filling) which, in the authors' opinion, is an important new development.

The authors' aims and aspirations will be fulfilled if they succeed in attracting the attention of theoreticians, experimenters and technologists to the unsolved problems of polymer filling, and in outlining some new and promising ways towards their solution.

References

1. Katz, H. S., Milewski, V. (ed.): Handbook of Fillers and Reinforcements for Plastics; New York: Van Nostrand Reinhold Co., 1978
2. Volfson, S. A.: Chemistry and Life (Rus.), N2, 16 (1984)
3. Nielsen, L. E.: Mechanical Properties of Polymers and Composites; New York: Marcel Dekker, 1974
4. Titow, W., Lanham, B.: Reinforced Thermoplastics; London: Applied Sci. Publ. 1975
5. Folkes, M. J.: Short-Fiber Reinforced Thermoplastics; New York: John Wiley, 1982

1. Principles of Polymer Reinforcement with Fillers

1.1 Matrix-to-Filler Stress Transmission Mechanisms

1.1.1 Reinforcing Fibers

It may be helpful to consider the mechanism of stress transmission in a composite, from the matrix to the filler, depending on the filler particle configuration.

Suppose the adhesion between the filler and the polymer matrix is strong, i.e. commensurate with the cohesion energy within the matrix.

Consider a composite reinforced with *continuous* fibers. A tensile stress is applied to a composite specimen along the fiber direction. This stress (σ_c) will be distributed between the matrix and fiber in the following manner:

$$\sigma_c = E_m v_m \varepsilon_c + E_f v_f \varepsilon_c \qquad \text{(I.1)}$$

where ε_c is the strain in the composite; E_m is Young's modulus of the matrix, E_f that of the fiber; v_m is the volume fraction of the matrix and v_f that of the fiber.

Note that in this case the relative elongation will be the same in both phases of the composite. Since $E_f \gg E_m$, at $v_f \approx v_m$ the portion of the stress transmitted to the fiber (σ_f) will be proportional to E_f/E_m.

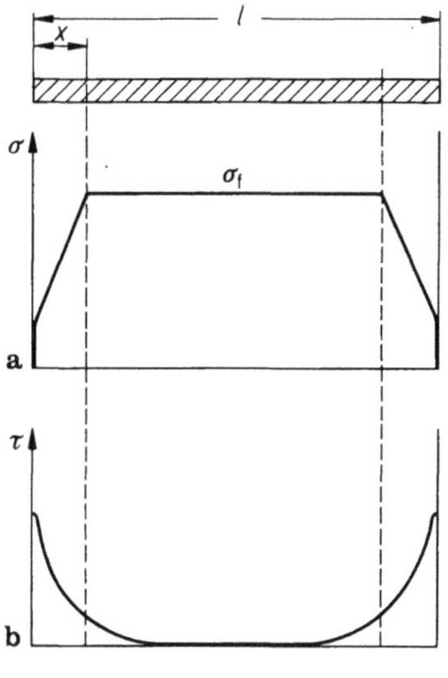

Fig. I.1. Composite reinforced with unidirectional, long discrete fibers: Diagrams of tensile stress (σ) applied to fiber of length l (**a**), and shear stress (τ) applied at the interface (**b**).

Now consider a composite reinforced with unidirectional *long discrete* fibers. The tensile stress is again applied along the fibers. Equation (I.1) is not applicable in this case since the stress will be unevenly distributed along each fiber. Figure I.1 shows schematically the tensile stress distribution along individual fibers (a) and the stress distribution at the interface (b).

The stress is transmitted to the fiber through the matrix. Assume that the fiber has a cross-sectional area S, radius r and length l. Consider an end segment of the fiber of length x. The mechanical equilibrium equation for this end segment is:

$$\sigma_x \pi r^2 = \int_0^x \tau(x)\, 2\pi R dx + \sigma_m \pi r^2 \tag{I.2}$$

where $\sigma_x \pi r^2$ is the tensile load transmitted to the end segment of length x by the rest of the fiber;

$\int_0^x \tau(x)\, 2\pi r dx$ is the portion of the tensile load transmitted by the tangential stresses developed at the interface to the side face of the fiber. Note that for an elastic matrix $\tau\ (x)$ is variable, whereas for a plastic matrix $\tau\ (x)$ may be assumed to be constant.

$\sigma_m \pi r^2$ is the portion of the tensile stress transmitted by normal stresses from the matrix to the fiber end face.

The stress at the fiber end is $\frac{\sigma_m \pi r^2}{\pi r^2} = \sigma_m$, that is, it is equal to the stress in the matrix.

The farther from the fiber end, the larger the value of the first term in Eq. (I.2); therefore the second term may be ignored for large x.

Is there any limit to the increase of stress applied to the fiber? Apparently the stress may grow until it becomes equal to that on a continuous fiber. This consideration enables us to define the so-called *critical fiber length* l_c. For $x \geqq l_c$ the stress transmitted to the fiber becomes maximum. In this case:

$$\sigma_x S_f = E_f \varepsilon_f S_f = \int_0^{x=l_c} \tau(x)\, 2\pi r dx \tag{I.3}$$

For a plastic matrix $\tau_x \simeq \tau_m$, hence

$$E_f \varepsilon_f S_f = \tau_m 2\pi r l_c \tag{I.4}$$

One can see that l_c is a function of ε_f, i.e. of the total stress applied to the composite specimen. The maximum stress that can be applied to a fiber is defined as its ultimate strength, that is:

$$(E_f \varepsilon_f)_{\max} = \sigma_f^* \tag{I.5}$$

Hence we obtain an equation for l_c under a critical load

$$l_c = \frac{\sigma_f^* r}{2\tau_m} \tag{I.6}$$

Since the critical fiber length is a function of the fiber diameter d, the relationship is usually written as

$$\frac{l_c}{d} = \frac{\sigma_f^*}{4\tau} \tag{I.7}$$

We have analyzed the situation arising at one end of the fiber. Taking into account both ends, the critical length given above must be doubled:

$$\frac{l_c}{d} = \frac{\sigma_f^*}{2\tau} \tag{I.8}$$

Comparison of the case of discrete fibers with that of continuous fibers shows that in the former case the end portions of the fiber will not experience the total load. This reduces the effective fiber length by the factor $l_c/2l$, at $l > l_c$ (see Fig. I.2)

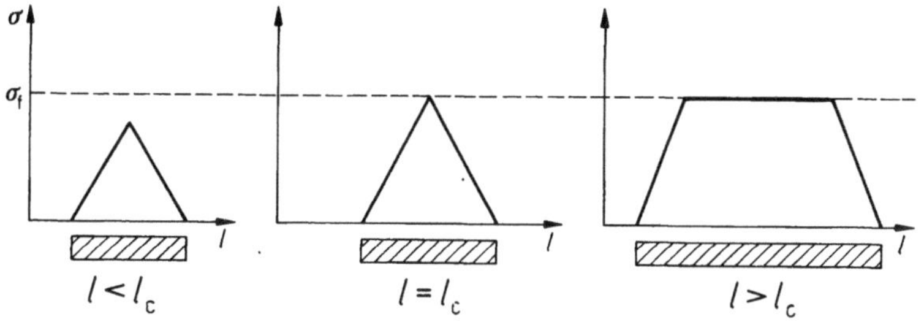

Fig. I.2. Diagram of tensile stress applied to discreet fibers of different lengths.

The shorter the filler fibers, the smaller is the portion of each fiber that effectively resists load in the composite. At $l = l_c$ the effective fraction of the fiber filler in the composite is only one half of the real fraction. This is illustrated by Fig. 1.2., where the area of the triangle is equal to half the value of $\sigma_f \cdot l_c$.

If the fiber length is below critical, it will not be fractured by whatever load is applied to the matrix. This means that a short fiber will not work to the full effect and the strength gain from the addition of such a filler will be low.

The maximum stress will occur at the middle section of the fiber, and for a plastic matrix it is (see Eq. I.3)

$$\sigma_{\max}\pi r^2 = \tau 2\pi r \frac{l}{2} \tag{I.9}$$

$$\sigma_{\max} = \frac{2\tau l}{d} \tag{I.10}$$

The average stress transmitted to the fiber of length $l \leq l_c$ is equal to half the maximum stress (see Fig. I.2):

$$\bar{\sigma}_f = \frac{1}{l} \int_0^l \sigma(x)\, dx = \frac{\tau l}{d} \tag{I.11}$$

Thus the reinforcement provided by very short fibers will be:

$$\sigma_c = \sigma_m v_m + \sigma_f v_f \tag{I.12}$$

At $l \simeq d$, (see Eq. I.11) the stress transmitted to a filler particle is equal to the shear stress τ generated in the matrix. Since $\tau_m = \sigma_m/2$, the strength of the composite will, in this case, be lower than that of the matrix. However, this conclusion does not take into account the transmission of stress by the filler particles' ends. For this reason the introduction of disperse fillers in moderate quantities does not reduce the composite strength.

In deriving the above equations we assumed that all the fibers in the composite specimen are oriented in the direction of load. Actually it is extremely difficult to prepare a specimen with an uniaxial orientation of short fibers. In cases where moldable thermoplastic-based or polyester-based composites are reinforced with short glass fibers, one has to face the problem of obtaining a material with properties as isotropic as possible, that is, with a random distribution of fiber orientations.

Even at $l \gg l_c$ the stresses transmitted to unoriented fibers are much lower than those transmitted to oriented fibers. In the simplest situations where one knows the fiber orientation in the composite one may characterize the effect by a reinforcement factor [1]. For uniaxial fiber orientation along the application of load the factor is apparently equal to unity. For a uniform statistical planar distribution it is $\frac{3}{8}$; for uniform statistical three-dimensional distribution it is $\frac{1}{6}$ [2].

In practice, especially in the case of moldable composites, the problem of strength design is much more complex. Due to the fact that some of the fibers are broken in the process of shaping, the fibers in the product have different lengths. Besides, the orientation pattern of the fibers is often extremely complicated and nonuniform in the volume of a specimen or finished part. That is why, for example, there is always a discrepancy between the strength characteristics of a test specimen and a finished part made of short-fiber-reinforced moldable thermoplastics.

1.1.2 Dispersed Fillers

From the above discussion it may be inferred that if a dispersed filler is used as reinforcement instead of fibers, the strength of the composite may decrease, so that the tensile strength may become even smaller than that of the unfilled polymer (i.e. the matrix).

This situation merits special discussion. Subjected to a tensile load, unfilled polymers are characterized by two critical points on the stress-strain diagram: the flow point and the ultimate stress. Dispersed fillers have only an insignificant effect on the flow limit (provided there is a good adhesion between the filler particles and the matrix), whereas the ultimate elongation (ε_c^*) becomes radically smaller.

The weak effect of the filler on the ultimate stress σ^*, may be attributed to the counterbalancing of two phenomena: stress concentration in the vicinity of the filler particles (consequently, reduction of the flow limit), and partial transmission of the stresses from the matrix to the filler.

The drop of ε_c^* in filled composites may be attributed to the smaller area occupied by the matrix in the cross-section of a composite specimen under tension. In this case one may expect an inverse proportionality between ε_c^* and the volume fraction of the filler. Such relationships are, indeed, experimentally observed for dispersion-filled elastomers. In plastics, however, the drop of ε_c^* is much sharper. It is quite illustrative that by adding as little as 10—15 vol.% of a filler to materials which unfilled have their ε values differing by an order of magnitude, the elongation may be reduced to about equal value.

This may be attributed to the development of cracks about the filler particles, and to the formation of particle agglomerates initiated by voids resulting from a local detachment of the matrix from the particles. If a filled specimen is subjected to a tensile load at an elevated temperature, the ε value increases because the material (like an elastomer) becomes less sensitive to flaws [3].

At high filling ratios the ε value may be expected to become smaller than that at the flow limit, and the composite will begin to fail at $\sigma < \sigma^*$. This means that the composite will undergo brittle fracture and it may be treated in terms of the theory of brittle (or quasi-brittle) failure.

According to this theory, first formulated by Griffith [4], the ultimate stress σ^* is given by

$$\sigma^* = A\sqrt{\frac{E\gamma}{c}} \tag{I.13}$$

where

A is a numerical factor of the order of unity;

E is the elasticity modulus of the composite;

γ is the effective surface energy of fracture, encompassing all energy dissipation processes accompanying crack propagation;

c is the effective defect initiating the development of the main crack.

At normal temperatures one observes brittle failure of filled thermoplastics already at filling ratios of 10–15 vol.%.

The addition of rigid filler particles to the composition causes an increase of the modulus of elasticity in proportion with the volume percentage of the filler. The effective surface fracture energy is higher in a composite than in an unfilled polymer for a number of reasons. Dispersed particles make the crack propagation path longer, absorb a portion of the energy, and enhance the plastic deformation of the matrix. Hence it follows from Eq. (I.6) that the strength of the composite should increase with the volume percentage of the filler. But this is not the case, because the value of c and the interaction between neighboring voids overweigh.

The value of c represents the size of the voids formed when the matrix becomes detached from the filler particles due to deformation. Naturally, the voids are the larger, the greater the size of the dispersed filler particles. From this we may conclude that it is plausible to use finely dispersed fillers that do not contain any large particles. Another important conclusion is that one may expect a considerable statistical scatter of measured strength data for composite samples, since a single void that has become critically large may initiate the main crack.

We have outlined the general behavior of a composite with a finely dispersed filler. Practice, however, shows that there are a number of complicating factors, viz.:

A. Matrix morphology in the composite

The most tangible effect of a filler on matrix morphology may be expected for semicrystalline polymers in which the filler will greatly affect the crystallization conditions. In certain cases, the experimentally observed reinforcement of dispersion-filled composites may be attributed to variations of matrix morphology.

B. Dispersed particles as additional crosslinks of the rubber-like network

This problem has been studied most thoroughly for elastomers. A considerable number of publications deals with this subject (see, e.g. [59–61]). It has been established experimentally that a disper ed filler has a reinforcing effect only if the filler particles are very small [5]. Obviously the filler particles are adsorbed by macromolecules and act as additional points of the macromolecular network. This results in a reinforcement of the material. However, it has not been found possible to explain this effect only by additional crosslinking.

C. Structure of dispersed filler particles

This effect has been most convincingly demonstrated for elastomers filled with carbon black. Most investigators agree that structurized filler particles hinder the development of cracks during straining of the material [5].

D. Porosity

High content of voids in filled systems is an inevitable side effect of their production process. Since voids initiate cracks under deformation, they reduce the strength of the composite.

The void content in many composites is high because of poor wettability of the filler particles by polymers. The presence or the release of water on the particle surface also promotes the formation of voids.

E. Agglomeration of dispersed filler particles

This phenomenon inevitably results in decreased mechanical strength of the composite due to an increase of the effective size of the filler particles and to the low strength of the agglomerates themselves.

F. Molecular restructuring of matrix

To process high-filled compositions, whose viscosity is accordingly high, one has to use elevated temperatures and shear stresses. These two factors promote degradation processes in the matrix, in particular a variation of the molecular weight distribution of the matrix during the shaping process. These effects should be taken into account if one compares the properties of filled and unfilled thermoplastics.

G. Residual stresses in composites

The coefficients of thermal expansion of polymers and mineral fillers differ by a factor of ten on the average. This fact is responsible for residual stresses remaining in the composite after curing. As regards the filler the stresses are compressive. Under a tensile load the compressive stresses should present an additional resistance to dewetting of the matrix from the filler, i.e. improve the composite strength. But in real materials, the existing structural inhomogeneities give rise to a complex stressed state, that is, at certain points shear and tensile stresses are present. This may facilitate crack propagation and ultimately cause reduction of the strength of the material.

1.1.3 Adhesion at the Interface

Special attention has been paid in the literature to the effect of adhesion at the interface in composites (see e.g. [62,63]).

In deriving equations relating the ultimate stress with the critical fiber length in Section 1.1.1, we implicitly assumed that the adhesion between the fiber and the polymer matrix is high. What does this mean?

In the case of "good" adhesion between matrix and fiber the maximum stress that can be transmitted from the matrix to the fiber is equal to the shear yield point of the matrix, τ_m, for a plastic matrix, and equal to the shear strength of the matrix for a brittle material.

In the case of "poor" adhesion the maximum stress transmissible from the matrix to the fiber will be smaller than τ_m and equal to the strength of adhesion.

From this, it seemingly follows that in complete absence of adhesion even a very small stress applied to the matrix would cause detachment of the matrix material from the fiber surface with formation of voids. No stresses at all would be transmitted to the fiber in this case.

Actually, however, the mechanism of stress transfer at the interface in composites is much more complex. One must distinguish between the normal stresses effective at the interface and shear stresses. The tensile stress is transmitted to the fibers by shear stresses along the fiber orientation. Normal stresses arise on the side surfaces of the fibers due to the residual thermal stresses in the composite and the different Poisson's ratios (μ) of the matrix and fiber (this value is usually higher for fibers than for polymer matrices). Both these factors lead to normal compressive stresses acting on the fibers.

Usually when we say adhesion we mean certain chemical or physical (electrostatic, dispersion, etc.) bonds across the interface. Since also the friction forces, arising under the effect of compressive stresses, mediate transmission of stresses from matrix to fiber, it may be more correct conceptually to use the term resistance to shear between matrix and fiber, (τ_{mf}), instead of adhesion strength. The former value can never be zero even when the adhesion is "poor". Although the case of no adhesion between the matrix and filler is treated in many theoretical works, this assumption is obviously hypothetical.

Unfortunately, there have been no reliable evaluations of the resistance to shear and the contributions of different factors to this value, and one is confined to qualitative statements.

In continuous fiber-reinforced composites the resistance to shear between matrix and fiber has a relatively small effect on the composite strength under tension along the fibers. In practice, it does have a certain effect at stresses close to the ultimate stress, when the "weaker" fibers begin to break and stress transmission through the matrix begins to play an important role.

Much more important is the effect of resistance to shear on such mechanical characteristics of composites as layer shear, transversal strength, compressive and flexural strength. Note that when tension is applied across the fibers it is not the resistance to shear but the resistance to separation (σ_{mf}) which begins to play the major role.

In the case of discrete fibers with l close to l_c, matrix-to-fiber stress transmission becomes important. If $\tau_{mf} < \tau_m$, we must use τ_{mf} instead of τ_m in the formula for l_c (see Eq. I.6), which will make l_c larger. The lower the τ_{mf} value the higher l_c for a given matrix.

At high ratios of filling with discrete unoriented fibers, the strength of reinforced composites will depend also on σ_{mf}. At low σ_{mf} it is easier for a matrix to come off the transversely oriented fibers and the resultant voids will initiate crack development. In composites of this type the σ_c versus v_f relationship always passes through a maximum. At low σ_{mf} the maximum values of σ_c will decrease.

As pointed out above, dispersion-filled composites do not have a high mechanical

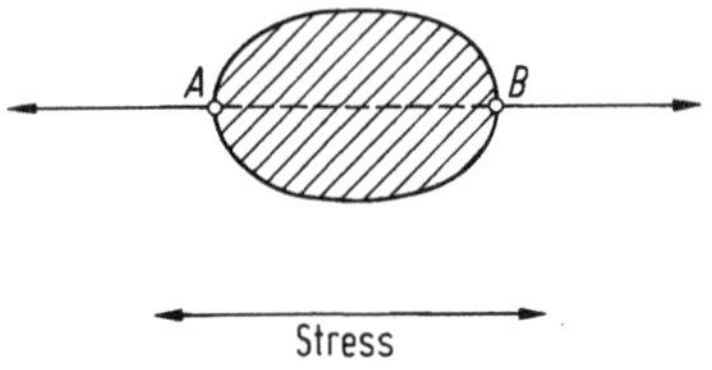

Fig. I.3. Diagram of normal str sses applied to a particle. A and B are the points in which normal stresses are maximum.

strength. Under tension, normal stresses are applied to each particle (Fig. I.3). The points at which such stresses are at maximum (points A and B in Fig. I.3) are the most vulnerable, since separation of the matrix from the filler is most likely there. Voids are formed which destroy the integrity of the composite and, as they grow and interact with each other, initiate cracks and promote specimen failure.

Obviously the higher the resistance to separation between the matrix and the filler, the higher the stresses that can be applied to the specimen before separation takes place.

Are the voids formed about the dispersed filler particles in a composite always undesirable and what is the risk involved? To answer this question we must consider three factors, viz. void size, void content and the stress causing dewetting.

The void size is determined by the filler particle size. The smaller the particles, the smaller will be the voids resulting from dewetting of the matrix material. The void content (the total volume of voids) is determined by the volume fraction of the filler in the composite. The stress required to cause dewetting depends on the adhesion strength between the filler particles and the matrix.

The stress at which a void becomes dangerous, i.e. begins to grow, is related with its initial size by a Griffith-type equation

$$\sigma \sim \frac{1}{\sqrt{p}} \tag{I.14}$$

where

p is the maximum void size, about equal to the particle size.

At higher filling ratios, the voids formed near individual particles begin to interact with each other. At very high filling ratios it becomes likely that the voids will merge and cause rapid specimen failure, as soon as the matrix begins to separate.

If a very good adhesion between matrix and filler is assured, that is, if the adhesion strength is higher than the matrix cohesion forces, then the σ_c value required to break the specimen will be controlled by the stress concentration in the vicinity of the filler particles.

What we have said above is valid only for brittle matrices, whereas in highly plastic ones the concentrated stresses will relax on account of the plastic flow of the matrix.

If the adhesion is very poor (i.e. the resistance to separation is low), voids will be formed already under low stresses, but this situation will not be dangerous for the structural integrity of the specimen. Such voids will, instead, be responsible for a pseudo-plastic behavior of the specimen. The strain/stress diagram of such a specimen will fully imitate that of an unfilled plastic polymer! It will have a "flow limit" at which

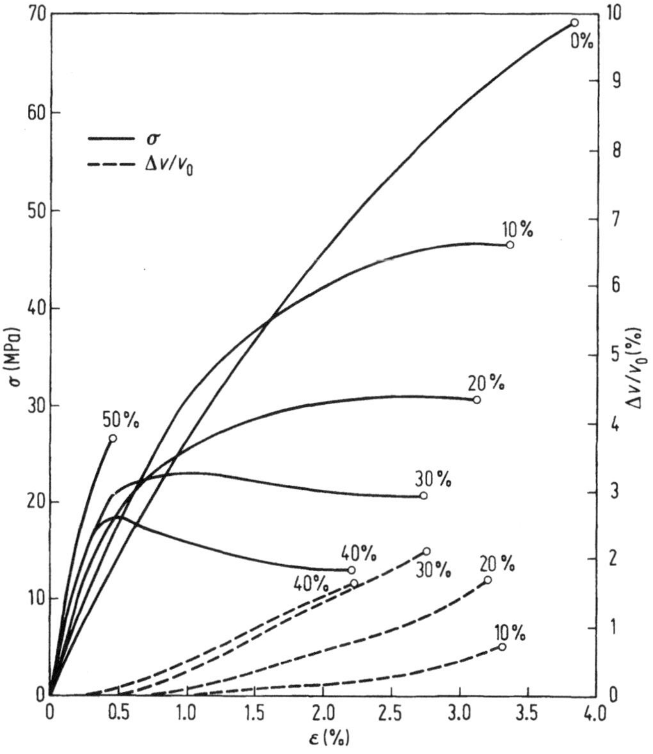

Fig. I.4. Stress-strain diagrams and volume increase with elongation for dispersed filled composite with poor adhesion between phases. Filler: glass beads, matrix: a polyurethane network [29].

the matrix will begin to separate from the filler particles, a "plastic flow region" under a relatively constant stress and sufficiently high relative elongation at rupture (Fig. I.4).

It is remarkable that similar behavior may be observed even in a composite whose matrix has no flow limit under the given experimental conditions. Rubber matrices are one example [6,29].

A principal difference between an unfilled polymer and a filled polymer with low adhesion between the phases is the concomitant volume variation. Plastic deformation cannot change the volume of an unfilled polymer since longitudinal elongation is compensated for by decrease of transverse dimensions. The Poisson's ratio in this case is 0.5. (Note that we are considering plastic deformation, for in case of elastic deformation the Poisson's ratio is 0.3 to 0.4.) In contrast, the volume of a dispersion-filled composite increases as the matrix separates from the filler particles (see Fig. I.4). In the limit, the increase of volume is proportional to the deformation, and Poisson's ratio becomes zero.

We now will consider the factors associated with particle characteristics: particle size, shapes and size distribution, and particle distribution over the matrix volume.

A. Effect of particle size

In the case of good adhesion between phases, stress concentration near the particles is independent of their size, and hence the σ_c versus particle size relationship is expected to manifest itself up to filling ratios of 0.4—0.5.

In the case of poor adhesion between phases, the moment at which the matrix actually separates from the particles depends only weakly on the particle size. But the subsequent history of a specimen will depend on it. For relatively large particles ($\simeq$100 μm), matrix dewetting will have a disastrous effect on the composite strength since the voids will immediately begin to grow.

B. Effect of filler particle shape

In the literature one often meets the statement that dispersed fillers with a highly developed surface are very effective [5,19]. This assumes *a priori* that the adhesion strength between fillers and matrix is very high. The presence of acute corners on filler particles should increase local stress concentration. On the other hand, the size of voids that may be formed at these points due to matrix dewetting may not be large.

Unfortunately, no systematic data on the effect of particle shape on composite strength have been published so far.

C. Effect of particle size distribution

As we have already noted, large-sized particles are undesirable when the adhesion between matrix and filler is poor. This is true for particles of varying size as well as for particles of the same size. Therefore, the general practice is to remove the coarse fraction of the filler before its use.

In the case of good adhesion the problem reduces to assure the maximum possible packing of particles within the composite. It is known that polydisperse particles may be packed closer than monodisperse ones. For a certain distribution function the maximum filling ratio may be close to unity.

The higher the maximum volume ratio of filling v_{max} (i.e. the closer the particles can be packed), the longer is the minimum distance between particles for a given filling ratio $v < v_{max}$. At v_{max}, filler particles contact each other, i.e. the minimum spacing of particles is zero. Thus for the same volume ratio of filling the minimum spacing between polydisperse particles will be smaller than between monodisperse particles.

The stress (strain) concentrations are at maximum precisely where the interlayers between the particles are thinnest. It is in these regions that the specimen failure begins.

Particular attention to the use of polydisperse fillers must be exercized when a high-filled composite is to be obtained. Important examples of such materials are solid jet fuels [7] and polymer-impregnated concrete [5].

1.1.4 Effect of the Matrix

Till now we have not considered in detail the role of the matrix in the reinforcement provided by filling. Whereas it is well known from practical experience that some polymers are susceptible to reinforcement, others are not, the requirements to filler particle size also depend upon the polymer matrix type.

For fiber-filled composites it follows from Eq. I.6 that the higher the flow limit of the matrix the lower the critical length of the fiber. This means, for example, that the fibers used to reinforce polyethylene have a critical length 4 to 5 times larger than those used for the reinforcement of nylon or polyacetal resins.

Moreover, it is known from practice, that different reinforcement effects are achieved by the addition of short glass fibers to different thermoplastics, even though the flow limits and Young's moduli of most commercial plastics are relatively close. Two reasons may be responsible for this:

1) Different adhesion strengths between matrices and fillers. The adhesion is usually the strongest in polar polymers (epoxy resins, polyesters, nylon, etc.), capable of forming H-bonds with the hydroxyls available on the glass fiber surface. By using specially selected substances named *coupling agents* it is possible to increase the adhesion between fibers and hydrophobic polymers.
2) The reinforcement effect depends largely on matrix ductility, i.e. the resistance to crack propagation. Thus, in order to obtain the maximum reinforcement effect it is necessary to use matrices with a high ductility. At high loading rates, e.g. in impact tests, high ductility values are particularly important. Most commercial plastics do not have a high ductility and, when filled to 40–60 %, become brittle and lose strength considerably. To improve ductility, technologists use such techniques as plasticization, elasticization, or reduction of the crystallinity of semicrystalline polymers by copolymerization.

Since the stresses are concentrated near the filler surfaces, and the processes responsible for initiation of fracture are also localized there, the question presents itself, whether it may be possible to modify the matrix at these zones by providing intermediate layers with special properties. There are two approaches to the problem:

1) The intermediate layer has a rigidity intermediate between those of filler and matrix, whereby stress concentrations may be relaxed [8,9].
2) The intermediate layer is highly elastic in order to prevent the propagation of voids formed when the matrix separates from the filler [10–12].

Note that, while the first approach appears to be attractive for filling with both, short fibers and dispersed particles, the second approach may prove effective only with dispersed fillers.

1.1.5 Conclusions

A. Fibers

For economic reasons glass fibers, both long and chopped, have become most important among fibrous fillers.

Strong, high-modulus organic fibers have found wide application in aeronautics and some other branches.

Mineral fibers and primarily asbestos, remain the traditional competitors with glass fibers (although the use of asbestos is increasingly discouraged because of its health hazards). Acicular Wollastonite crystals and basalt fibers have also been used commercially. Attempts have been and are being made to grow acicular mineral crystals. Metallic and organic fibers find application. Wood flour, which has a fibrous structure is the traditional filler for phenolic and aminoformaldehyde resins.

The highest reinforcement effect is, apparently, obtained from carbon fibers. It may be interesting to mention that carbon fibers themselves are not at all stronger than glass fibers, but their modulus of elasticity is 3–5 times higher. The higher Young's modulus of composites filled with carbon fibers, compared with those filled with glass fibers (at the same concentration), leads to a situation where the fraction of load transmitted to the fiber is higher, whereas that received by the matrix is lower [13].

The elasticity modulus of asbestos fibers is about 1.5 times higher than that of glass fibers. But these fibers have a very uneven length distribution which reduces the reinforcement effect. Moreover, the small diameter of these fibers hampers the processing of asbestos-filled compositions.

Compared to glass fibers, standard organic (polymer) fibers have a lower Young's modulus (by about a factor of 10) and provide a correspondingly lower reinforcement effect.

Natural organic fibers, including cellulose, wood flour, etc., provide almost zero reinforcement effect due to their small size and great flexibility.

The KEVLAR fiber has proven to be an excellent continuous fiber. Due to a high Young's modulus and high specific strength KEVLAR might be expected to be also a superb short-fiber filler. So far there have been few reports of its use in this form [14]. The high anisotropy and low flexural modulus of this fiber may limit its range of application.

B. Lamellar Fillers

Apart from fibrous fillers, composites may be reinforced with lamellar fillers. For these fillers the aspect ratio is defined as the ratio of linear dimension to thickness.

It is known that owing to their form lamellar composite fillers exhibit reinforcing effect in two directions, rather than in one as in the case of fibers. Therefore, other conditions being the same (same strength, size much in excess of the critical one), randomly distributed lamellar fillers may be expected to produce a better reinforcement effect than fibers, for the same filling ratio. Experimental data for special grades of mica [15,16] confirm this expectation (glass fibers were used as reference). Of course, it is extremely difficult to obtain mica with a given aspect ratio and maintain it through the blending and shaping processes.

Plates can be more densely packed in a composite than fibers. Therefore, high volume contents of a lamellar filler may be obtained. For example, with mica as filler one may obtain materials with a Young's modulus 2–3 times higher than that of glass fiber-reinforced plastics.

Such common fillers as talc and kaolin have also a lamellar structure, but do not exhibit a reinforcing effect due to their low aspect ratios.

C. Dispersed Fillers

Since, as we have mentioned above (Section 1.1.2), dispersed fillers do not improve the strength of a composite, we now have to consider whether they reduce the strength.

Brittle fillers of little strength, such as perlite, will fail in a loaded specimen. This will initiate cracks and an early specimen failure.

Organic fillers may be divided into two groups, elastic and pseudo-plastic ones. The first group has been extensively used for improving the impact strength of materials (impact polystyrene, ABC-plastic, impact PVC, etc.). In all cases the added elastomer reduces, however, the tensile strength of the composite (compare polystyrene and impact polystyrene).

The second group includes fillers prepared from plastic wastes, nutshells, wood flour, etc. Typical examples of plastic wastes are powdered thermoset plastics and powdered polymer mixtures. Their effect on the strength of the composites depends on the composition of the plastic wastes and on the strength of the polymer matrix. As usual, the strength of the composites is reduced compared to the pure polymers, although this negative effect may be offset by an improvement of other properties.

1.2 True and Effective Deformation of Polymer Matrix Composites under Tension

It is well known that unfilled polymers behave in either of two characteristic ways in tensile tests:

a) brittle fracture of specimens at relatively small strains ($\leqq 10\%$);
b) ductile failure of specimens preceded by formation of a "bottleneck" at high deformations (tens or hundreds %).

It is commonly believed that "brittle" plastics (polystyrene, poly(methyl methacrylate), cross-linked epoxy and polyester resins, etc.) cannot withstand considerable strains. Actually, this is not so, and under specific conditions all these materials behave very plastically. For instance, even epoxy resins with a high density of cross-linking, prepared as thin films, have been found in tensile tests to have a true strain of 40% at the mouth portion of the main crack [17]. Deformation of polystyrene and poly(methyl methacrylate) leads to the formation of crazes, i.e. cracks whose surfaces are bridged by highly stretched polymer. The true deformation of polymer in these bridges amounts to about 100%.

In macroscopic specimens of the above-mentioned materials it is impossible to realize the potential plasticity margin because of their low resistance to cracking.

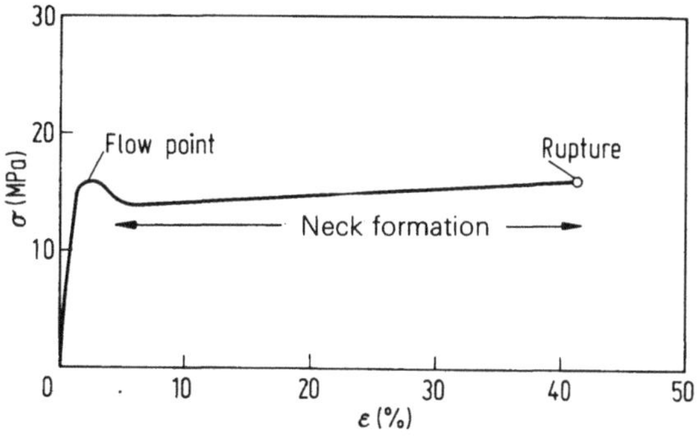

Fig. I.5. Typical stress/strain diagram for impact polystyrene.

If some way could be found to increase their resistance to cracking, their deformability would sharply increase.

Impact polystyrene is an illustrative example for such effort. The addition of dispersed rubber particles to a polystyrene matrix sharply increases the cracking resistance of the composite, and under tension it exhibits ductile behavior with necking [18].

The relative elongation at rupture of the specimens of different commercial grades of impact polystyrene is 25–50%. It is important to stress that this considerable elongation is not at all due to the rubber filler but to the deformation of the continuous polystyrene phase itself! The rubber content is usually low, within 5 to 10%.

The neck formation begins, in a stretched sample, after the maximum of the σ–ε diagram, which determines the flow limit (cf. Fig. I.5). The material is gradually entrained in the neck section and becomes oriented and reinforced there. In this case, sample straining is nonuniform, and the mean strain ($\bar{\varepsilon}$) reflects the true deformation in the neck (ε_n) and the portion of the material in the neck section (h):

$$\bar{\varepsilon} = \varepsilon_n \cdot h \tag{I.15}$$

In the limiting case all the material passes into the neck, followed by rupture. Then, $\bar{\varepsilon} = \varepsilon_n$. In other cases fracture may occur before all material passes into the neck. Then $\bar{\varepsilon}$ will not reflect the true strain of the material. The presence of some kind of defects may be the reason of such premature failure. In other words, the $\bar{\varepsilon}$ value measured in this case will be spurious and cannot be regarded as a characteristic of material deformability.

In brittle plastics to which a rigid dispersed filler has been added, the fracture under tensile loads depends on the matrix-filler adhesion strength. At high strength of adhesion between the phases the material will fail in a brittle manner. However, when the adhesion is not strong, crazes may be formed in the matrix under tension which will manifest themselves by a flexure point (flow limit) and pseudo-plasticity in the σ–ε curve.

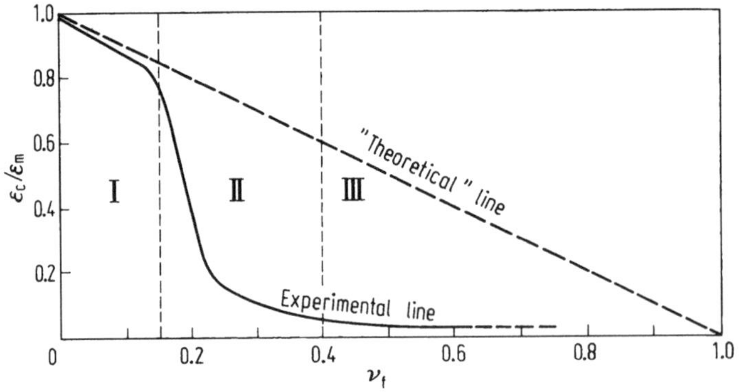

Fig. I.6. Diagram $\varepsilon = f(\nu_f)$ for dispersed filled composite (ductile polymer matrix). Region I is the ductile failure. Region II is the intermediate mechanism. Region III is the brittle failure.

A complex situation arises when a rigid dispersed filler is added to a ductile polymer matrix. What effect will a filler of this kind have on matrix deformation?

The volume ratio of the polymer in the composite specimen will be lower. In the simplest case the deformability of the matrix polymer itself will be unaffected and

$$\varepsilon_c \approx \varepsilon_m^* v_m \qquad (E_f \gg E_m) \tag{I.16}$$

As we have already mentioned (Section 1.1.2), in practice one observes a much sharper drop of $\varepsilon_c = f(v_f)$ than one would expect theoretically (Fig. I.6). Analysis of the experimental data reported in the literature for ductile polymers allows to single out three regions in the $\varepsilon_c = f(v_f)$curve:

Region I corresponds to a small filler content and insignificant variation of ε_c compared with the unfilled polymer; region III corresponds to a high filler content, about equal to close packing. Brittle behavior is unconditionally dominant in this region; region II is an intermediate zone and is of special interest. Characteristically, the data reported for composites falling into this region (containing 15–40% of filler) are the most contradictory, and the reproducibility of the results is poor. Some samples have high elongations, while others have low ones.

Analyzing Fig. I.6 one has to answer two questions: 1) why is the elongation at break lower than predicted by the theoretical curve; 2) what is the reason for the instability of the results in the transition region II.

Equation (I.16) assumes that only the matrix is deformed and the deformation is homogeneous throughout the volume. In fact, this is not the case. Matrix deformation is inhomogeneous, being especially large in between filler particles. The existence of such inhomogeneity means that in case of contacting filler particles there may appear regions, where breaking is initiated at low average matrix deformations. The number of these regions sharply increases with the increase of v_f. It would seem that the critical v_f values should be close to the most dense packing of the filler particles [13]. Nevertheless, from Fig. I.6 it can be seen that embrittlement occurs at much lower v_f values (~ 0.15). One may suggest that this is caused by the formation of a loose three-dimensional network of filler aggregates in the matrix. Several computations made by Manevich and Oshmian seem to confirm this concept [65].

Unstable results obtained from the deformation of samples of region II can be explained from the analysis of the samples themselves. Figure I.7 presents 3 samples

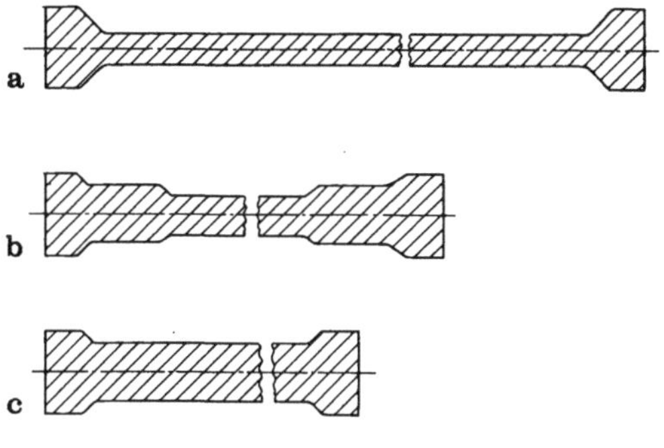

Fig. I.7. Test samples of dispersed filled HDPE; a: the whole material passed into the neck; b: part of the material passed into the neck; c: brittle failure.

of HDPE filled with chalk ($v_f = 0.2$), where for the same v_f the ε values corresponded to ductile, brittle, and intermediate failure. In one case the composite behaved in the same manner as unfilled HDPE, with all of the material passing into the neck (Fig. I.7a), with an ε value of 150%.

The brittle failure of the other composite sample (Fig. I.7c) with the same filler content ($v_f = 0.2$) at $\varepsilon = 8\%$ can be attributed to local filler concentration fluctuations, for in a number of experiments the local concentration at the point of rupture was about 1.5 times the average v_f value.

Specimen b in Fig. I.7, which failed at $\varepsilon - 63\%$, is particularly interesting. It is clearly seen that necking had begun as the load was applied but it was not complete before the specimen finally failed. In this case, too, analysis of the fracture point revealed an unduly high local filler concentration. Judging by the specimen appearance one may conclude that the premature failure of this specimen was due to internal flaws and that the measured ε value is an apparent rather than the true one.

Generally, the location of region II and the value of ε will depend on the filler particle size. In another series of tests carried out under carefully controlled conditions we obtained HDPE matrix composites with tufa as the filler ($v_f = 0.22$). The mean particle size of the filler was varied. The experimental results are given in the table below.

Table 1.1. Composite of HDPE and tufa; ε value as a function of the mean particle size

particle size, μm	v_f, %	ε, %
2	22	200
10	22	120
30	22	15

The systematic variation of ε with the mean filler particle size, as shown in Table 1.1, is plausible in view of the above discussion. It may be safely assumed that, as the filler particle size increases, the effective flaw size also increases, i.e., the specimen becomes more defective and the probability of brittle (or intermediate) failure increases.

There are data showing that in some cases even very fine fillers ($\simeq 2$ μm) may increase composite brittleness. Apparently, this is caused by agglomeration of filler particles or increase of the effective size of agglomerates.

In summary, from the above data one may formulate the hypothesis that in filled ductile matrices the transition from plastic to brittle behavior occurs in a jump. The "spreading" of this transition, with the formation of the transition zone II on the $\varepsilon = f(v_f)$ curve, is due primarily to filler concentration fluctuations. Another cause for the formation of Region II may be the aggregation of filler particles, or the presence of coarse fractions of filler.

1.3 Fracture Toughness and Impact Strength of Polymer Composites

Most specialists in the field assume that the addition of solid (or even gaseous!) fillers to plastic matrices makes the latter more brittle, if the filler content (by volume) is around or above 0.2. The problem of reducing brittleness is a very urgent one, since the concomitant reduction of the impact strength of high-filled plastics often limits their applicability considerably.

At the same time, a class of so-called "impact-resistant" composite materials, is widely known consisting of plastics filled with dispersed elastomeric particles. The most popular are impact polystyrene, ABS-plastic, impact (frost-resistant) polypropylene, impact PVC, and others.

The problem of rendering materials more impact-resistant has long been attracting the attention of scientists. A comprehensive report in this field has been published by Bucknall in 1977 [18]. Some theoretical findings, and to an even greater degree the recent experimental achievements, have stimulated attempts to introduce elastomeric phases into composites with solid fillers to improve their impact strength.

Impact testing of materials is a complex task. There are currently at least two dozens of experimental procedures [19] but none of them gives sufficient information to substantiate conclusions about the material performance in service.

According to the definition given by Phillips and Harris [20], fracture toughness must be regarded as a complex of properties. All contributions making up what is called fracture toughness have one thing in common: they refer to the work spent for causing material failure. The work is determined by the energy absorbed by the specimen at different stages of fracture. But the relative contributions at the different stages depend strongly on the size and geometry of the specimens.

For example, in tests using notched specimens (the so-called Charpy and Izod technique), the cracking resistance of the material is highly important. In notchless specimens, the ability of the material to initiate cracks is paramount. Thus, materials reinforced with discrete fibers often show high resistance when tested as notched specimens, but have a lower strength than the matrix polymer when tested without notching.

In this section we will consider the effect of composition and structure of polymer matrix composites o the fracture toughness. Table 1.2 shows which co tributions

Table 1.2. Contributions to fracture toughness, as a function of the test specimen's geometry and the rate of deformation

Rate of deformation	Test specimen	
	Notched	Without notch
Low	Surface energy of fracture	Work of fracture
High	Impact strength	Impact strength

to fractur toughness can be experimentally established depending on the sample configuration and deformation rate.
The surface energy of fracture, γ, depends on tensile strength σ and notch depth c according to:

$$\gamma = \frac{\pi\sigma^2 c}{2E} \tag{I.17}$$

The work of fracture is defined by the area underneath the stress-strain curve.

The different components of fracture toughness are interrelated. The relationships have been unambiguously established for brittle homogeneous and isotropic materials for which they have been evaluated quantitatively; they are less unambiguous for plastics and polymer materials, due to the contributions from visco-elastic and plastic deformation.

As regards filled composites, the parameters of fracture are largely determined by the properties of the polymer matrix, since most of the energy required to fracture the material is used for straining and fracturing of the polymer matrix. But since the mechanism of failure of composites is different from that of unfilled matrices, it is impossible to calculate the fracture toughness of a composite from the individual data for matrix and filler.

The introduction of a dispersed filler increases the surface energy of material fracture. But it must be specified that this is only true for brittle matrices such as epoxy and polyester resins [21,22]. In non-brittle polymer matrices, the filler reduces this parameter due to the decrease of the volume fraction of matrix in the plastic zone. For polyamides the specific surface energy of fracture is 10^3–10^4 J/m^2, whereas for brittle cured resins it is of the order of 10^2 J/m^2.

In brittle matrix composites the increase of surface energy of fracture has practically no effect on impact strength, which systematically decreases as the volume ratio of filler increases.

Most practitioners in this field believe that in ductile polymers there is a correlation between the work of fracture and the impact strength; concerning the dispersion-filled composites it is assumed that there is a correlation between the reduction of the relative elongation at rupture and the decrease of the impact strength.

There is also the perception that in composites of this type (e.g. with nylon as the matrix) a decrease of adhesion between matrix and filler causes both an increase of the relative elongation at rupture and a certain increase of impact strength. However, this effect is achieved at the expense of a lower ultimate stress.

A comparison of pure and glass-filled commercial polymers shows that they can be divided into two groups according to their impact strength behavior:

1) Glass fibers reduce the impact strength of polymers that feature a high impact strength when unfilled (polycarbonate, nylon, ABS-plastic);
2) Glass fibers have little effect or even improve the impact strength of polymers whose Izod impact strength is below 100 J/m and which are very sensitive to notching (polypropylene, polyacetals, polyesters, polyphenylene oxide).

This phenomenon may be attributed to the fact that glass fibers perform a two-fold function in polymer matrix composites: they absorb the extra energy required

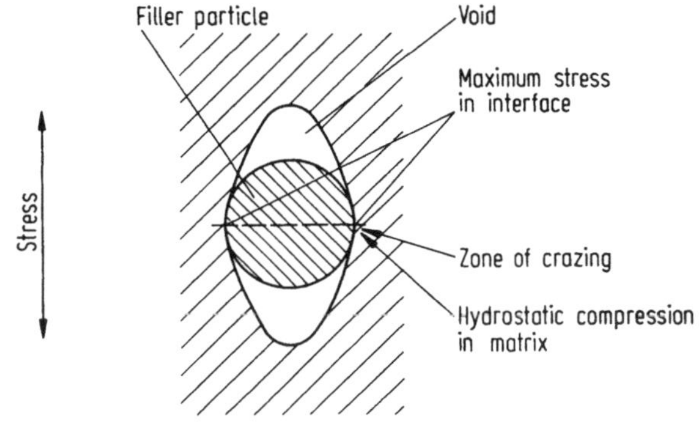

Fig. I.8. Diagram of the deformation of a composite filled with disper ed rigid particles.

to pull them out of the matrix during crack propagation; and there is increased stress concentration in the matrix near the fiber ends.

In brittle polymers the size of the plastic zone at the crack mouth is smaller than, or commensurate with, the fiber spacing. Therefore, in the case of medium filling ratios (up to 40 wt %), the polymer matrix shows little sensitivity to the effects at the fiber ends.

In ductile polymers the size of the plastic zone is large and encompasses the fiber ends. Therefore, the matrix becomes brittle rapidly as the stress concentration increases.

Figure I.8 illustrates the mechanism by which a slight increase of impact strength may be caused by dispersed rigid particles, in cases where the adhesion between the dispersed phase and the matrix is small. The matrix separates at earlier stages of deformation (at relatively low loads); on further increase of the load the maximum tensile stresses will be localized near the equator of the particle and will be directed parallel to its surface. Depending on matrix properties and loading conditions this will cause the appearence of crazes, shear strips or shear flow. All these changes will increase the fracture toughness of the composite, although the presence of hydrostatic compression tends to prevent the appearence of crazes. Accordingly, the increase of fracture toughness is small.

If the adhesion is good, a high load is neccessary to cause separation of the matrix at the points where the stress concentration is maximum (on the interface), which results in the formation of a void initiating crack development. Consequently crazes, shear strips and shear flow do not originate and fracture toughness does not increase.

1.3.1 Effect of Spatial Distribution of Filler on Impact Strength

The effect of the spatial distribution of particles is one of the principal problems in the relationship between structure and composite properties. It has been treated theoretically in many papers and monographs. The basic qualitative conclusion is that the distribution must be as uniform as possible and mixing must be as thorough as possible. Otherwise, the localities with a higher-than-average filler content will be potential sources of brittle failure.

The first step in a quantitative treatment of this problem was to search for a quantitative description of the spatial distribution of particles in a composite. The papers of Tovmasyan et al. [23-27] were pioneering in this respect. These authors proposed to consider the distribution of particles in agglomerates. The term agglomerate implies all those particles separated from each other by a distance smaller than a certain r_0. The weak point of this approach is the arbitrary selection of the r_0 value. However, there is reason to hope that further research in this field will help to attach a certain physical meaning to this parameter.

A better definition of this parameter may be based either on the analysis of strain and stress fields, or on the measurements of the strain-strength characteristics as a function of the filling ratio. But already at its present stage such analysis may be of interest. The authors suggest to consider the moments of the function describing the distribution of the number of particles in the agglomerates, as characteristics of the spatial distribution of the filler. The first moment represents the average number of particles per agglomerate. The higher moments (especially the third and fourth) are particularly sensitive to agglomeration. The spatial distribution characteristics are then used as a means of analyzing the technology-structure-properties relationships [24-27].

The experimental investigations were carried out with HDPE filled with glass balls, aluminum hydrate, glass fibers, etc. Besides, mathematical structure simulation methods were used which consisted of stochastically distributing particles with a given size distribution function and a given volume fraction. The model structures constructed in this manner and experimental microphotographs of low-temperature fracture surfaces in notched composite specimens were then processed in the same way.

In a separate experiment it was shown that such a fracture surface reflects the particle distribution in the bulk of the sample. After that the test specimens obtained by different processes (pressing, mixing with different intensities followed by molding, etc.) were compared with random "model" specimen. It was found that pressing cannot assure a uniform distribution of particles: the third and fourth moments are unduly high in comparison with the model sample. (The volume fraction of the filler is the same in all cases, of course.) The particle distribution in molded samples is sensitive to the mixing operation and ideally may come close to the model distribution. However, there is some difference between distribution patterns in and across the molding plane, the distribution being close to that of the model in that plane and markedly different in the perpendicular plane. It is hard to tell which distribution may be expected in a steady-state polymer flow. We believe that finding this out is an interesting theoretical and experimental problem. The first rough results reported in the work under discussion seem to corroborate this point of view.

Analysis of the mathematical model for different particle size distributions has shown that the degree of agglomeration is the smaller, the broader is the size range. Thus, a purely mathematical experiment has confirmed the plausibility of using fillers with a broad range of particle sizes, even if the composite under consideration contains only a small amount of filler.

A preliminary analysis of the relationship between the spatial distribution of the filler and certain physical and mechanical properties of the composite has also been made, though for a rather narrow size range. It has been found that both the elasticity modulus and flow limit of the composite were, up to filling ratios of 15%, little sensitive to the values of the distribution moments, whereas the impact strength considerably increased when the distribution was made more uniform (see Table 1.3). In general terms, this has been known since time. In other words, brittle matrices (higher loading rates are analogous to a more brittle matrix) are especially sensitive to the spatial distribution of the filler: agglomeration of particles promotes the initiation of a brittle crack, ultimately causing the failure of the specimen. On the basis

Table 1.3. Impact strength, a, and ultimate stress, σ^*, as a function of the uniformity of the filler distribution; HDPE filled with glass beads (13.5% [27])

Specimen	No. of particles in the picture	Moments of spatial distribution of particles M_1	M_2	M_3	M_4	σ^* (MPa)	a kg cm/cm^2
Extrusion on the BRABENDER plastograph	55	1.6	2.6	4.7	6.8	43 ± 3	23 ± 2
Statistical model	55	1.3	1.6	2.1	2.7	—	—
ΔM		0.3	1.0	2.6	4.1	—	—
Double extrusion on the BRABENDER plastograph	50	1.5	2.2	3.4	4.7	44 ± 3	33 ± 4
Statistical model	50	1.2	1.4	1.9	2.4	—	—
ΔM		0.3	0.8	1.5	2.3	—	—

Table 1.4. Impact strength, a, as a function of the spatial distribution of particles [27]

Composition	Part of the distribution	ΔM_1	ΔM_2	ΔM_3	ΔM_4	a, kg cm/cm^2
PE + 25% of finished glass beads	Left	0.2	0.6	1.4	1.8	12.6
	Right	0.2	0.4	0.7	1.0	18.5
PE + 15% glass beads	Left	0.2	0.6	1.8	3.3	17.5
	Right	0.2	0.7	1.4	2.0	23.1

ΔM_i is the difference between the moments of spatial distribution of the test specimen and that of the corresponding model.

of this approach, quantitative relationships between structure and properties may be established.

It is important that this approach has not only been helpful in establishing the relationship between the averaged parameters of spatial distribution of filler particles and averaged impact strength values for a given series of composite specimens, but also in determining the correlation for specimens of one and the same series.

Comparison of the impact strength data and moments of spatial distribution for the best and worst specimens (right and left part of the distribution function of the impact strength) has shown that, in this case too, there is a correlation (Table 1.4).

1.3.2 Elastification of Rigid Matrices by Filler

The stress-strain diagram at room temperature (Fig. 1.9) shows that the addition of elastomeric particles to a brittle amorphous matrix results in flow behavior and a considerable increase of the relative elongation at rupture (i.e., the work of fracture).

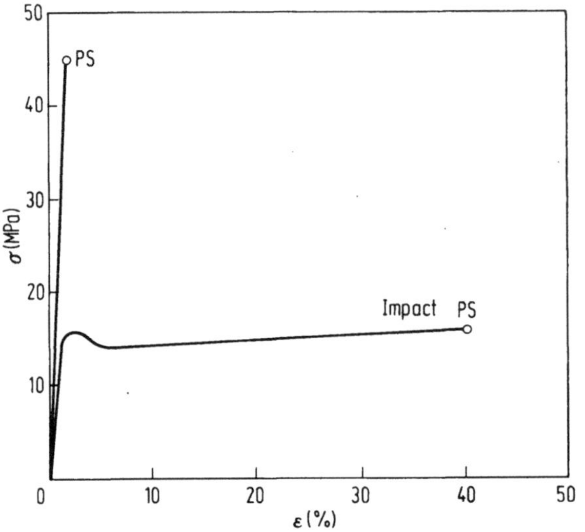

Fig. I.9. Stress/strain diagrams for polystyrene and impact polystyrene [18].

But a similar effect has been obtained by the addition of glass beads with an average particle size of 30 μm to a brittle matrix. For the first time this was demonstrated for a styrene-acrylonitril copolymer matrix [28]. In a more recent publication on this subject [29] it has been shown that, for loosely cross-linked polyurethane filled with glass beads, the flow limit on the stress-strain diagram and whitening of the specimens were due to a separation of the filler from the matrix, crazing and formation of voids.

It should be stressed, however, that the addition of glass beads leads to an increase of the ultimate tensile strain from 1.5–2 to 3–6%, but not to an increase of the impact strength.

For the present discussion, the following factors are important:

a) Mechanism of increase of impact strength
b) Variations of the σ vs. ε diagram
c) Effect of heterogeneous system morphology

a) Reinforcement mechanism

Newman in his review on rubber modified plastics [10] has reported on six reinforcement theories, indicating that neither provides a satisfactory explanation of the observed experimental facts.

After Newman and Strella [30] had shown that the energy absorbed by the rubber particles is only 10% of the total energy absorbed by the composite, it became clear that the part played by the rubber particles is very complex. These authors advanced a hypothesis according to which the following factors are considered as critical:

1) The deformation of the polymer matrix around the rubber particles involves an increase of the free volume;
2) the rubber particles can relax under the effect of combined stresses;
3) the rubber particles can arrest a catastrophical crack growth.

Later, Bucknall and Smith [31], Kambour [32], and others attributed the whitening of rubber-modified polystyrene under tensile stress to the formation of multiple crazes. This phenomenon underlies Bucknall's theory of multiple crazing of glassy polymers [33].

A craze is a zone in a strongly plastically deformed porous material which has the form of a crack whose sides are bridged by fibrils of stretched polymer material. This distinguishes a craze from a microcrack or a crack. The bridging fibrils ensure stability of dimensions and high strength of the material.

The problem of crazing and the detailed structure of crazes has been thoroughly studied by Rabinowitz and Beardmore [34], Kambour [35] and, more recently, in a book edited by Kausch [58]. The formation of crazes has been convincingly demonstrated by different techniques in glassy linear and loosely crosslinked (network) polymers, e.g., polystyrene and its copolymers with acrylonitrile and methyl methacrylate, poly-(vinyl chloride), polycarbonate, etc. Reports have appeared about crazing in semi-crystalline polymers (polypropylene [35]) and densely cross-linked network polymers (epoxides [36]). In the latter case, however, the proofs are not as convincing; many authors believe that crazing can only occur in a material capable of withstanding high local strains (> 100%).

As noted above, crazing under tension is not an exclusive privilege of rubber particles. The same effect has been observed in glass bead filled plastics [29]. Figures I.8 and I.10 illustrate a principal difference between a rigid and an rubberlike filler.

A rubber particle (having a good adhesion to the matrix due to a block-copolymer grafted onto its surface) is deformed under stress without volume variation. The stress concentration is highest along its equator, and the hydrostatic tension there helps the craze formation (Fig. I.10); matrix separation is absent.

In the case of glass beads, crazes are formed only if there is no adhesion between filler and matrix [29]. The strain pattern around a rigid particle is shown in Fig. I.8. Crazing is possible, but with a much lower probability than in the case of elastic particles owing to the hydrostatic compression. Simultaneously, a crack begins to grow as a result of matrix separation and void formation.

According to Bucknall [18], the role of rubber particles is not limited to crazing. The ability of rubber particles to deform due to the extra stress at the growing craze mouth also plays a part. Rigid particles do not behave in this way.

Formation of crazes cannot account for the entire picture observed in impact tests

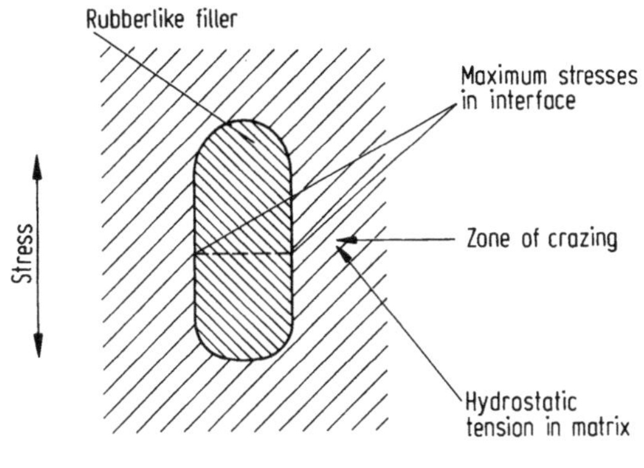

Fig. I.10. Diagram of the deformation of a composite filled with dispersed rubberlike particles.

of rubber-modified polymers. Besides, one always observes a shear flow which sometimes manifests itself as shear strips. Bucknall and Street [37] estimated the relative contribution of crazing and shear flow in the creep behavior of various polymers and concluded that the influence of either may vary in a wide range depending on the material properties and experimental conditions.

We are not aware of any such analysis of impact data of materials filled with rigid particles. It may be hypothesized, however, that some shear flow must occur even in these materials. In case of no adhesion between phases, the matrix constraint will be determined solely by particle spacing, as with rubber particles. In case of strong adhesion, the matrix constraint is much more severe. Apparently, even if the adhesion is not so good, shear deformations do not have time to take shape before a catastrophic opening of cracks, initiated by the voids formed as a result of matrix separation.

Bragaw [38] suggested that, in the case of rubber particles, a part of the impact energy is absorbed by craze branching. This mechanism is specific to elastic fillers and should not manifest itself with a rigid filler.

b) The stress-strain diagram

Mixing rubber particles into brittle polymer matrices, such as polystyrene, PAN, PMMA, results in the appearance of a flow limit and flow region on the stress-strain diagram. According to Bucknall [18], the appearance of the upper flow limit in impact polystyrene is not due to the variation of the specimen cross-section (as it is in the case of necking), but is due to intensive crazing and a reduction of the density (increase of specimen volume).

The situation is the same for glass beads which normally have a poor adhesion to the matrix. It is important to stress again that for rigid fillers with good adhesion no such phenomenon is observed. In the case of rubber particles, the adhesion to the matrix has, apparently, a small effect on flow limit and relative elongation at rupture.

Addition of rubber to ductile polymers such as polyamide, PVC, polycarbonate, and polypropylene results in an extension of the flow range and a decrease of the upper flow limit and strength. We may remind that ductile polymers containing rigid fillers fail in a brittle way and the behavior of the upper flow limit depends on the strength of adhesion. Again the behavior of rubber-filled systems is simulated by composites containing glass beads with a low adhesion to the matrix. For poly(phenylene oxide) filled with glass beads it was found that the upper flow limit decreases as the filling ratio increases. However, if an interphase layer is used to provide high adhesion, this relationship is quite different and the upper limit variations are small [39].

The effect of temperature on the behavior of filled polymers deserves special attention. Bucknall [18] studied this relationship for rubber-filled polystyrene and showed that the temperature dependence of the notched Izod impact strength is determined by the glass transition point, Tg, of the rubber. The impact strength of the polystyrene matrix remains practically unchanged, as long as the T_g of polystyrene (92 °C) is beyond the experimental temperature range. A tendency for an increase of the impact strength with temperature is observed with this and other amorphous and semicrystalline polymers after passing over the T_g. This tendency is seen also with solid fillers, since the impact strength of the material is determined by the matrix behavior only in this case.

According to Petrich [40], the addition of an impact strength modifier to PVC has a similar effect as the increase of temperature. The probability of brittle failure is reduced due to the lowering of the flow limit.

c) Effect of heterogeneous system morphology

The first rubber-modified plastics were polystyrene compounds with coarse-particle rubber. The relatively large lumps of rubber (up to 100 µm) of irregular shape increased the impact strength only insignificantly even when present in fractions of up to 20–30%.

The use of polymerization methods for the synthesis of impact polystyrene and ABS-plastics has brought about at least four radical changes in the dispersed phase morphology:

1) decrease of the particle size to 0.1 to 10 µm and the possibility of controlling both the average size and the size distribution;
2) provision of a grafted block-copolymer layer assuring a good inter-phase adhesion;
3) complication of the internal structure of the rubber particles due to polymer matrix occlusion;
4) partial cross-linking inside the rubber particles which allows their elastic modulus, shape and resistance to shear to be controlled.

The combination of these factors underlies the superiority of the polymerization method over compounding, as regards the properties of the final product, primarily the combination of a high impact strength with high deformability and strength. Many works dealing with the problem of improving the morphological parameters of polystyrene plastics have been published [10,18,41], but even more are kept secret as manufacturer's know-how.

The experience collected in the work with polystyrene plastics has proven very useful for the modification of other polymers. It is well known that there is an optimum size and optimum rubber phase concentration for each polymer matrix. Summarizing some experimental works, Bucknall [33] has noted that the lower critical particle size boundary is a particularly important characteristic. For impact polystyrene it is 0.8 µm, for ABS and PVC 0.2 µm. The more ductile the matrix, the smaller is the critical particle size limit. It is generally assumed that the impact strength gradually decreases with increased particle size, but no extensive data are available thus far. Our own experience with impact polystyrene [41] shows that the impact strength of composites sharply decreases even if the rubber particle size is increased only from 10 to 20 µm.

Apparently, the particle size distribution and the average spacing between particles are also important factors, but no reliable data have been reported.

The lower critical size limit of rubber particles is commensurate with the medium size of rigid fillers.

All authors agree that a transition layer having a good adhesion both to the matrix polymer and to the rubber phase is necessary. It may consist of a graft or linear block-copolymer, partly miscible with each phase. For a polystyrene matrix styrene/butadiene rubber copolymers are suitable materials; for PVC partially miscible elastomers are used, such as chlorinated polyethylene [42], or a terpolymer of methyl methacrylate, butadiene and styrene, or an ethylene — vinyl acetate copolymer [43].

The complication of the morphology of the dispersed rubber phase is particularly marked with polystyrene matrices, and is due to the phase transitions during the

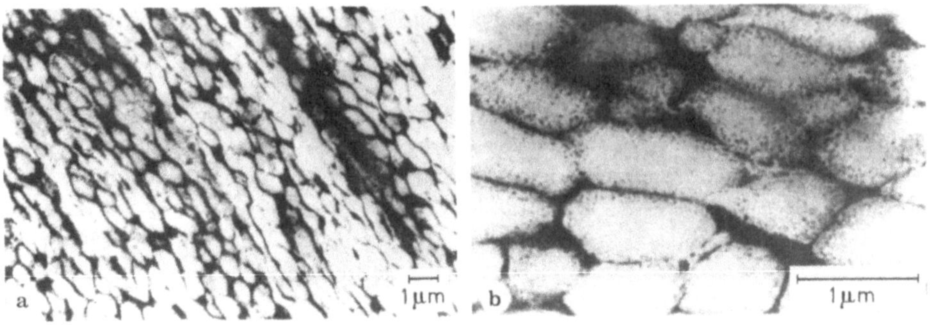

Fig. I.11. The honeycomb structure of the continuous rubber phase in a PVC/rubber composite [43].

polymerization. Normally, the content of polystyrene occluded within the rubber particles is 60–90% of the volume of the dispersed phase. The properties of such mixed particles extracted from impact polystyrene have been studied separately [44]. It is not the structure of these particles per se which is responsible for the high impact strength; only when dispersed in the brittle matrix they exhibit this effect.

In the contrasting case of rubber reinforced PVC, it is the honeycomb structure of a continuous rubber phase in which the PVC particles are dispersed, which assures improved resilience of the composite (Fig. I.11).

The two above-mentioned structures are, apparently, the two possible extremes of impact-resistant polymer systems. A rigid mineral filler is perfectly compatible with the structure of the first type, if the filler particles are coated with rubber.

Partial cross-linking in the bulk of the rubber particles is typical of polystyrene composites. When the degree of crosslinking is low, there is a high probability of deformation and even rupture of the rubber particles by the shear forces during processing. A denser cross-linking increases the elastic modulus of the particles, rendering their behavior similar to that of rigid particles. Newman [10] has described the structure of rubber particles used for improving the impact strength of other amorphous glassy polymers. In these particles the core was loosely cross-linked, and the shell was glassy.

However, there are no indications that the elastomer additives mixed into PVC and PP for improving impact properties are subjected to partial cross-linking.

1.3.3 Elastomer Shells Around Solid Particles

It may suffice to take a look at an EM picture of impact polystyrene (see Fig. I.12) to get the idea that an elastic shell might be just what is needed to impart impact resistance to composites filled with solid particles or fibers. Newman [10] discussed this question in his review of rubber-modified plastics.

The effect of interphase layers on the properties of composites filled with spherical particles and short fibers has been treated theoretically by Broutman and Agarwal [8]. They concluded that when the elastic modulus of the interphase layer is about one tenth of that of the matrix, one may expect the greatest reduction of stress concentration in the matrix and, indirectly, an improvement of the impact strength. Similar

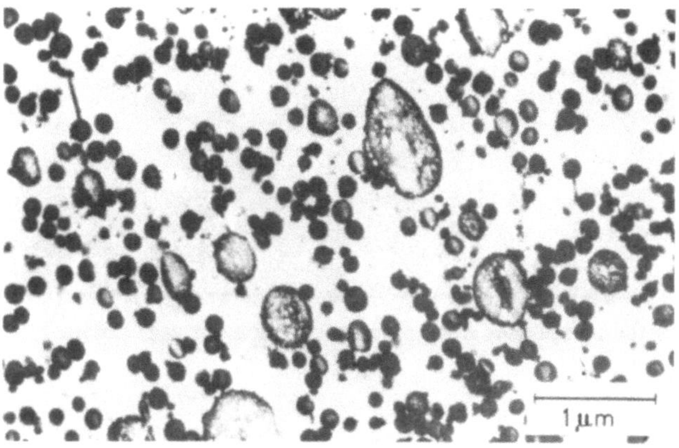

Fig. 1.12. Electron micrograph of an ultrathin section of ABS resin [66].

calculations have been given later [9] but, unfortunately do not permit to predict the resilience of composites.

A few patents have been issued, (more on them in Ch. III), in which various ways of providing solid particles or fibers with resilient coats are disclosed. However, no reliable experimental data are given.

The modification of chalk surfaces with rubber prior to adding the chalk particles to impact polystyrene has been reported [45]. In laboratory tests, the authors were able to encapsulate practically all particles. However, the resultant increase of impact strength was much lower than expected. Even a superficial analysis of the results helps reveal at least two reasons of the ill-success:

1) The modified filler particles did not contain an intermediate layer of grafted styrene/rubber copolymer, without which it was only natural that the composite should behave poorly;
2) large filler particles of >20 µm, even encapsulated, may produce a drastically negative effect on the impact characteristics of the composite.

Kitamura [46] has reported a considerable increase of the impact resistance of composites comprising polypropylene mixed with calcium carbonate or talcum, if the fillers were enclosed with an ethylene/propylene block-copolymer (EPR) and polyethylene (PE). Schematically, the structures of the composites obtained by the patented process [47] are shown in Fig. I.13.

The strength of adhesion between the polypropylene matrix and the EPR rubber and polyethylene in the composites has been measured [48]. The Izod impact strength and the adhesion strength multiplied by the PE domain surface area were found to be in good correlation. The authors assume that the impact energy is absorbed either due to stratification of the PE domains in the EPR interface, or due to void formation in the PE domains. If this is really so, this mechanism is quite different from the mechanism of failure of EPR-modified unfilled PP [49].

DOW CHEMICAL has reported [50], that the addition of an impact modifier, viz. chlorinated PE, to rigid PVC, together with the filler (chalk in quantities of 50 to 100 wt.

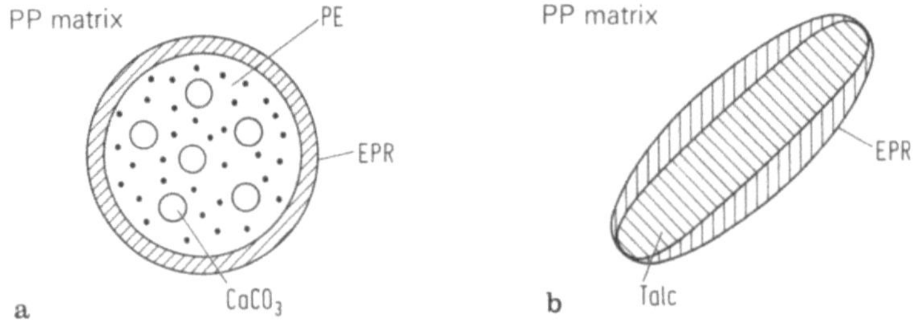

Fig. I.13. Schematical structure of the composites if the fillers were enclosed with polymer[47].

parts per 100 wt. parts of PVC) improves the impact strength of the composite by as much as a factor of 3–4. The recommended modifier concentrations (15–20% of PVC weight) are high and seem to indicate a honeycomb structure of the composite. In such a structure microcracks are formed predominantly in the PVC domains enclosed in the elastomer phase, but lack of experimental data does not permit an argumented conclusion to be made.

1.3.4 Alternative Ways of Impact Strength Improvement: Lowering of the T_g of the Matrix and Grafting of a Polymer to the Filler

Attractive as the elasticization technique would seem, the difficulties involved in controlling the particle morphology apparently stimulate research for other impact strength improving techniques.

One of the possible ways is to increase the matrix resilience by copolymerization, producing an effect similar to that of elevated temperature. As noted earlier, one may expect an appreciable increase of impact strength only around the T_g of the matrix, or its amorphous phase. Therefore, in practice we can expect an improvement of the impact strength only for semicrystalline polymers such as polypropylene, polyformaldehyde, nylon, etc.

It should be noted that many of the experimental results concerning the impact improvement of polypropylene by filling with dispersed solid particles[51] may actually be attributed to the modification of the crystalline structure of this material.

Although to-date most industrially used filled compositions have been based on commercial grade polymers, the future will hopefully see the development of new matrix materials designed specifically for the use in composites.

An alternative technique consists in forming a graft polymer layer on the filler particle surfaces. The existing preparative and commercial processes for providing such layers will be discussed in detail in Ch. III.

There are at least two approaches to what we have called *polymerization filling*:

a) The grafted layer is of the same composition and average molecular mass as the matrix polymer, thus improving the compatibility at the interface.
b) The grafted layer has a different composition and/or average molecular mass from the matrix polymer.

The realization of the first approach may be exemplified by attempts to graft polystyrene onto short glass fibers [52] and PVC onto chalk particles [53]. The modified glass fibers were used as reinforcement of ABS-plastic. According to reports of MITSUBISHI MONSANTO, the addition of as little as 20% of such fibers increased the Izod impact strength by as much as 50% compared to unreinforced material. The ultimate tensile strength and Young's modulus were also reported to rise. These facts appear to indicate that the reinforcement mechanism is, in this case, different from the one active in the case of elastomers.

Chalk with PVC grafted onto its surface was used as modified filler for PVC. It has been reported that the impact strength of the compound increased when the

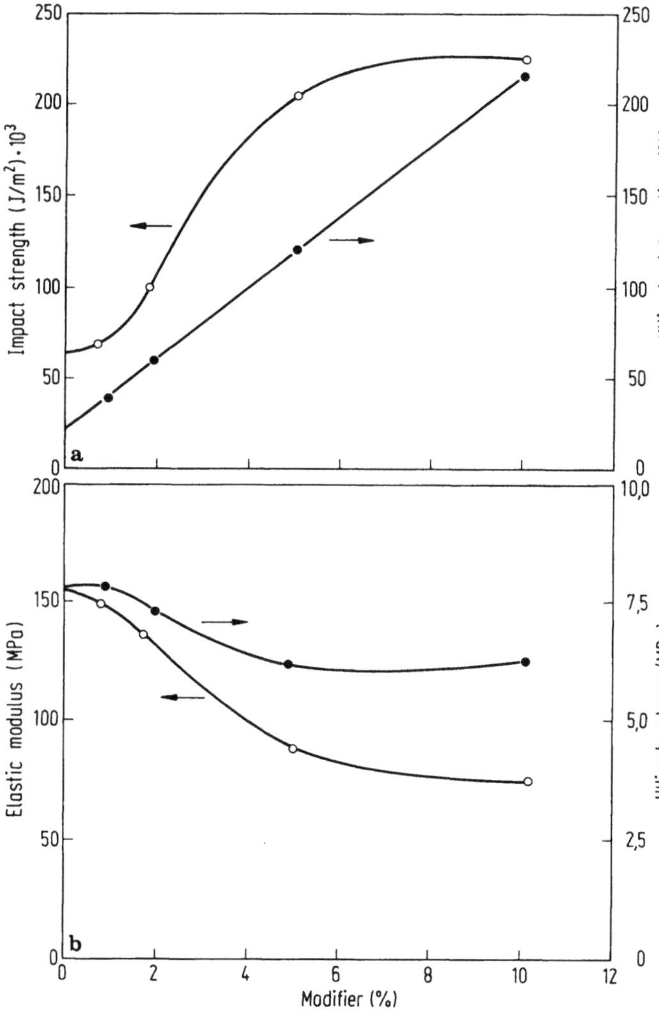

Fig. I.14. Effect of chalk modification with liquid oligomeric ethylene oxide on the mechanical properties of a LDPE-composite [54].

grafted polymer content was about 5% of the filler weight. Unfortunately none of the publications concerned with polymeric coupling agents discussed the mechanism of reinforcement.

The possible ranges of the composition and molecular mass of the interphase layer grafted to or adsorbed by the filler are really immense. At one extreme is the work of Polish investigators [54] on chalk modification with liquid oligomeric ethylene oxide. The chalk was used for filling LDPE. The effect of filling on the relative elongation to rupture, ultimate stress, elastic modulus and impact strength is illustrated in Fig. I.14. The authors note that in tensile tests the specimens withstood complete necking even with filler contents of 50 wt%, and conclude that the liquid modifier apparently arrests crack initiation effectively.

In later works [55,56] it was shown that similar results are obtained for chalk-, talc- and kaolin-filled HDPE. The effect cannot be explained by plastification of the matrix, nor by morphological changes or reduced crystallinity. The impact strength increased with filler and oligomer contents, beginning from a chalk concentration of 20 wt% (in comparison with the non-modified filler). These examples correspond to the case, where no adhesion and friction between filler and matrix are present (see Chap. 2.6).

On the other end of the line we may place the work by Enikolopyan et al. [57] where ultra high-molecular weight PE (UHMWPE) in quantities of 5 to 20% was grafted on to tufa, calcite, kaolin, and perlite particles, and the resultant filler was mixed with HDPE and LDPE by compounding (Fig. I.15). The impact strength of the compound as a function of the filler content had a maximum at 50–60% of filler. The ultimate tensile strength varied similarly. A "normal" dependence of Young's modulus on the filler concentration as well as a low relative elongation at rupture indicate that the

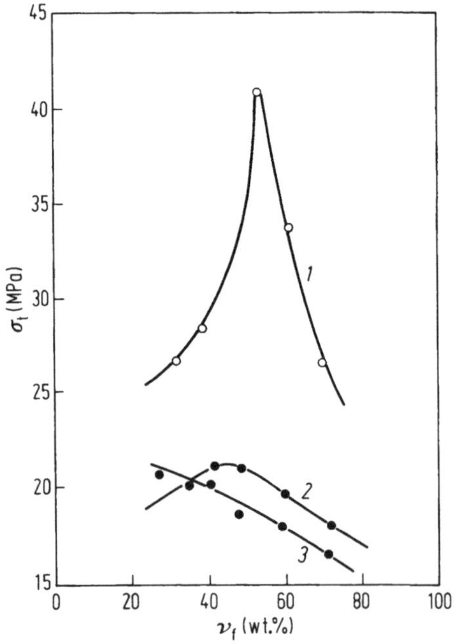

Fig. I.15. Dependence of tensile strength σ_t of composites HDPE-tufa on the weight fraction of filler ν_f (size of the particles is 8 μm). *1*: 20% UHMWPE grafted onto tufa and blended with HDPE; *2*: tufa, UHMWPE (20% to filler) blended with HDPE; *3*: tufa blended with HDPE.

reinforcement mechanism in this case is totally different from the one we discussed previously (for liquid cover around filler). The ultra high-molecular weight PE may be responsible for the high strength level of the composite.

1.3.5 Conclusions

We have described in much detail the state of the art concerning impact plastics modified with elastomers because the behavior of these materials has been studied most thoroughly. No doubt that the knowledge accumulated will stimulate further work, the ultimate aim of which will be to undermine the current conception of a solid particle filled polymer as a fragile material.

But much ingenuity will still be needed to arrive at a reasonable theoretical interpretation of the experimental results. There are serious grounds to believe that the reinforcement mechanisms active in rubber-filled systems will be applicable to solid particle filled composites as well.

References

1. Krenchel, H.: Fiber Reinforcement, Copenhagen; Akademisk Forlag 1964
2. Nelsen, L. E.: J. Appl. Polym. Sci. *10*, 97 (1966)
3. Smith, T. L.: Trans. Soc. Rheol. *3*, 113 (1959)
4. Griffith, A. A.: Phil. Trans. Roy. Soc. (London) A221, 163 (1920)
5. Manson, J. A., Sperling, L. H.: Polymer Blends and Composites, New York: Plenum, 1976
6. Takemori, M. T.: Polym. Eng. Sci. *18*, 1193 (1978)
7. Delmonte, J.: Ch. 10 in Handbook of Fillers and Reinforcements for Plastics, Katz, H. S., Milewski, V. (Eds.): Van Nostrand Reinhold Co., New York, 1978
8. Broutman, L. J., Agarwal, B. D.: Polym. Eng. Sci. *14*, 581 (1974)
9. Arridge, R. G. C.: ibid. *15*, 757 (1975)
10. Newman, S.: Ch. 13 in Polymer Blends, Paul, D. R., Newman, S. (Eds.), New York: Academic Press, 1978
11. Lavengood, R. E., Michno, M. J.: Proc. Div. Techn. Conf., Engrs. Props. and Structure. Div. Soc. Plast. Engrs, p. 125 (1975)
12. Trostyanskaya, E. B., et al.: Plast. Massy No 6, 50 (1976)
13. Katz, H. S., Milewski, (ed.): Handbook of Fillers and Reinforcements for Plastics; New York: Van Nostrand Reinhold Co., 1978
14. Sturgen, D. L., Lasy, R.: Ch. 27 in Ref. 13.
15. Woodhams, R. T., Xanthos, M.: Ch. 19 in Ref. 13.
16. Maine, F. W., Shepperd, P. D.: Compos. *5*, 193 (1974)
17. Pakhomova, L. K. et al.: Vysokomol. Soed., A23, 400 (1981)
18. Bucknall, C. B.: Toughened Plastics, London: Applied Sci., 1977
19. Nielsen, L. E.: Mechanical Properties of Polymers and Composites, vol. 2, New York: Marcel Dekker, 1974
20. Phillips, D. C., Harris, B.: Ch. 2 in Engineering Polymer Composites, Richardson, M. (Ed.), London: Applied Sci., 1977
21. Lange, F. F., Radford, K. C.: J. Mater. Sci. *6*, 1197 (1971)
22. Hammond, J. C., Quayle, D. C.: Proc. 2nd Int. Conf. on Yield Deformation and Fracture of Polymers, Cambridge, 1973
23. Tovmasyan, Yu. M. et al.: Dokl. AN SSSR *270*, 649 (1983)
24. Tovmasyan, Yu. M. et al.: Theses of VIth Int. Symp. Polym. Compos., Budapest, April 1983, p. 32
25. Tovmasyan, Yu. M. et al.: Theses of Vth Nat. Conf. Compos. Mater., Moscow, Oct. 1981, v. II, p. 198

26. Tovmasyan, Yu. M. et al. ibid. v. II, p. 200
27. Tovmasyan, Yu. M. et al.: Reaction Kinetics and Mechanisms, v. 5, Chernogolovka (Moscow District) 1982, p. 56
28. Lavengood, R. E., Nicolais, L., Napkis, M.: J. Appl. Polym. Sci. *17*, 1173 (1973)
29. Yilmazer, U., Farris, R. J.: Polym. Compos. *4*, 1 (1983)
30. Newman, S., Strella, S.: J. Appl. Polym. Sci. *9*, 2297 (1965)
31. Bucknall, C. B., Smith, R. R.: Polymer *6*, 437 (1965)
32. Kambour, R. P.: Macromol. Rev. *7*, 134 (1971)
33. Bucknall, C. B.: Ch. 14 in Ref. 10
34. Rabinowitz, S., Beardmore, P.: CRC Crit. Revs. Macromol. Sci. *1*, 1 (1972)
35. Kambour, R. P.: J; Polym. Sci. *D7*, 1 (1973)
36. Lilley, J., Holloway, D.: Phil. Mag. *28*, 215 (1973)
37. Bucknall, C. B., Street, D. G.: SCI Monogr. *26*, 272 (1967)
38. Bragaw, C. G.: Amer. Chem. Soc. Adv. Chem. Ser., *99*, 86 (1971)
39. Truss, R. W., Chadwick, G. A.: J. Mater. Sci. *11*, 111 (1976)
40. Petrich, R. P.: Polym. Eng, Sci. *13*, 248 (1973)
41. Malkin, A. A., Volfson, S. A.: Polystyrene, Moscow: Khimiya, 1975
42. Schramm, J. N., Blanchard, R. R.: Proc. SPERETEC, Cherry Hill, New Jersey (Oct. 1970)
43. Michler, G., Gruber, K.: Plaste und Kuatsch. *23*, 346 (1976)
44. Wagner, E. R., Robeson, L. M.: Rubber Chem. Technol. *42*, 641 (1970)
45. Academy of Science of GDR: Technical Communication 1983
46. Kitamura, H.: Progress in Science and Engineering of Composites, Ed. by Hayashi, T., Kawata, K., Umekawa, S., ICCM-IV, Tokyo, p. 1787 (1982)
47. Kokai, J. P., J. P. Presentation No. Sho-57-23642
48. Kuroda, T. et al.: Polym. Prepr. Japan, *25*(5), 893 (1976)
49. Speri, W. M., Patrick, G. R.: Polym. Eng. Sci. *15*, 668 (1975)
50. Dow Chemical Corp.: Techn. Information AEG80 (1980)
51. Novokshonova, L. A., Fridman, M. L.: Unpublished results.
52. Egbers, R. I.: Plast. World *37* (10), 66 (1979)
53. Grishin, A. N., Gusev, V. V. et al.: Plast. Massy No 9, 15 (1983)
54. Galeski, A., Kalinski, R.: In: Polymer Blends: Processing, Morphology and Properties, New York, Plenum, 1980, p. 435
55. Kalinski, R., Galeski, A., Kryszewski, M.: J. Appl. Polym. Sci. *26*, 4047 (1981)
56. Bardan, B., Galeski, A., Kryszewski, M.: ibid. *27*, 3669 (1982)
57. Fridman, M. L., Enikolopyan, N. S. et al.: Proc. IX Symp. Synthesis a. Props, Polym.-Filled Polyolefins, Chernogolovka (Moscow District) 113 (1982).
58. Crazing in Polymers, H. H. Kausch, ed., Advances in Polymer Sci., *52/53*, Springer-Verlag, Berlin—Heidelberg, 1983
59. Kraus, G.: Fortschr. Hochpolym.-Forsch., *8*, 156 (1971)
60. Vold, M. J.: J. Colloid Sci., *18*, 684 (1963)
61. Medalia, A. I.: J. Colloid Interface Sci., *32*, 115 (1970)
62. Plueddmann, E. P. (Ed.): Interfaces in Polymer Matrix Composites. New York: Academic Press 1974
63. Plueddmann, E. P.: Silane Coupling Agents, London: Plenum Press 1982
64. Donatelli, A. A., Thomas, D. A., and Sperling, L. H.: Recent Advances in Polymer Blends, Grafts, and Blocks, Sperling, L. H. ed., N. Y.: Plenum 1974, p. 375
65. Manevich, L. I., Oshmian, V. G.: Dokl. Akad. Nauk SSSR, 1985 in press.
66. Matcuo, U.: Polym. Eng. Sci., *9*, 206 (1969)

2. Theoretical Models for Describing the Physical and Mechanical Properties of Polymer Composites

The present state of the theory aimed at describing the physical and mechanical properties of filled polymers is such that some of the properties can be calculated to within a sufficiently high accuracy; only very approximate solutions are available for another group of properties, while the remaining properties can be found only from experiments.

The aspects of small deformations have been developed best in terms of the elasticity theory for brittle, plastic, and elastic polymer matrices and model systems of the fillers.

The behavior of composites under heavy deformations is more complicated, because the mechanisms of heavy deformations are rather diverse and depend much on the preparation and testing conditions, even for the same type of composite.

The mechanisms defining ultimate strength under tensile, compressive, bending, impact, and fatigue loads have been least developed up to now.

.The problems outlined above have been discussed qualitatively in the previous chapter. In this chapter we give a brief description of the available theoretical approaches to these problems, for polymers filled with particles and with short or continuous fibres. The chapter will also give information necessary to calculate some physical and thermophysical properties of polymer composites.

2.1 Calculations of the Elastic Properties of Dispersely Filled Composites

The simplest problems of the mechanics of composite materials involve the calculation of the physical and physico-mechanical properties of a material showing linear elastic behavior, i.e. proportionality between small deformations and applied stresses. It is important to remember that the deformations should be small not only on the average over the entire material, but also at any point of a specimen to be analyzed.

Dewey [1)] was the first to accurately calculate the shear modulus for an elastic medium with a small* volume fraction of spherical inclusions.

---

* In theoretical calculations, a "small" volume fraction means that filler particle interactions can be neglected. In practice this is the case for $v_{fl} < 0.1$.

The shear modulus equation is of the following form [1]:

$$\frac{G_c}{G_m} = 1 - \frac{15(1-\mu_m)\left[1-\frac{G_{fl}}{G_m}\right]\cdot v_{fl}}{7-5\mu_m+2(4-5\mu_m)\left(\frac{G_{fl}}{G_m}\right)} \qquad \text{(II.1)}$$

where
G is the shear modulus, the indices c, m, and fl refer to the composite, matrix, and filler, respectively; μ_m is the Poisson ratio of the matrix, v_{fl} is the volume fraction of the dispersed filler.

The special case of absolutely stiff inclusions ($G_{fl}/G_m \gg 1$), in an incompressible matrix ($\mu_m = 1/2$) is of interest and importance from a practical viewpoint. In this case expression (II.1) is transformed to the simple form

$$\frac{G_c}{G_m} = 1 + 5/2\, v_{fl} \qquad \text{(II.1')}$$

When calculating Young's modulus, (modulus for uniaxial tension, E) and the Poisson ratio μ of an isotropic material, the expressions

$$E = \frac{9KG}{3K+G} \quad \text{(II-2)}; \qquad \mu = \frac{3K-2G}{2(3K+G)} \qquad \text{(II.2')}$$

relate E and μ to the calculated parameters G (shear modulus) and K (bulk modulus) [51].

The solution of the problem of an elastic medium with a small bulk fraction of elastic filler gives the following equation for the bulk modulus, in a general form [1]:

$$K_c = K_m + \frac{(K_{fl}-K_m)\, v_{fl}}{1+\left[(K_{fl}-K_m)\Big/\left(K_m+\frac{4}{3}G_m\right)\right]} \qquad \text{(II.3)}$$

Thus, Eqs. (II.1), (II.2), (II.2'), and (II.3) make it possible to calculate the elastic parameters of a composite with a small quantity of spherical filler. It should be noted that Eqs. (II.1) and (II.3) include only elastic parameters of matrix and filler and are independent of geometric factors (particle size, filler distribution in the matrix volume).

It should also be noted that the equations of solid body linear elasticity and Newtonian viscous flow show some similarity, restricted however to the case of absolutely stiff inclusions which corresponds to the flow of solid particle suspensions. Accordingly, Eq. (II.1') is analogous to the diluted suspension viscosity expression obtained by A. Einstein in 1905.

$$\frac{\eta}{\eta_m} = 1 + 5/2\, v_{fl} \qquad \text{(II.4)}$$

The transition from small to medium and high volume fractions of fillers, i.e. from non-interacting to interacting inclusions, gives rise to considerable theoretical difficulties. In this case the geometric parameters of a particular medium, namely the filler particle size distribution and arrangement, have to be taken into account. As a result, exact general solutions have not been obtained for higher filler concentrations. Instead, various concrete models of a more or less special nature have been analyzed by now.

In the theory of composites two types of models are usually discussed:

1. Regular models where filler particles of the same spherical or rectangular shape are arranged to form a regular (cubic, hexagonal, etc.) lattice. In this case, the linear elasticity problems have been solved, either numerically (finite element method, etc.) [2,3,4], or analytically, in some special cases.
2. Stochastic models where, however, definite simplifying assumptions are made regarding the character of the size and space distribution of the particles. The so-called polydisperse model (proposed by Hashin [5]) is one example. In this model it is supposed that each filler particle of radius *a* is surrounded by a matrix layer, and that the size distribution of the resultant two-layered particles of radius *b* is such that the space is filled completely. The a/b ratio is assumed to be constant for particles of any radius. It should, certainly, be borne in mind that here we deal with a special case of simultaneous space and size distribution of particles. A good agreement of the model calculations with experimental data may be expected only if the filler particle size distribution is sufficiently broad and the components are properly mixed.

A similar, though simpler model (from the viewpoint of obtaining the end result), the three-phase model, was proposed by Kerner [6] and Van der Pool [7] and was examined thoroughly by Smith [8] and by Christensen and Lo [9]. This model assumes that a spherical filler particle is surrounded first by a matrix layer and then by a certain equivalent medium showing the properties of the composite.

Both models are adequate in that they permit composite properties to be analyzed within the entire range of filler concentrations (from 0% to 100% by volume).

According to the polydisperse model, the effective bulk modulus is given by

$$\frac{K_c - K_m}{K_{fl} - K_m} = \frac{v_{fl}}{1 + \left[(1 - v_{fl})\,(K_{fl} - K_m)\Big/\left(K_m + \frac{4}{3}G_m\right)\right]} \tag{II.5}$$

The polydisperse model does not permit to obtain a general shear modulus expression for the entire range of filler concentrations. At small volume fractions of filler, an expression identical with Eq. (II.1) is valid; for high volume fractions the following equation was given by Hashin [5]:

$$\frac{G_c}{G_{fl}} = 1 - \frac{\left(1 - \frac{G_m}{G_{fl}}\right)\left[7 - 5\mu_m + 2(4 - 5\mu_m)\frac{G_{fl}}{G_m}\right]}{15(1 - \mu_m)}(1 - v_{fl}) \tag{II.6}$$

(the terms of higher order in $(1 - v_{fl})$ are reglected here).

It is important to note that the bulk modulus expressions obtained in terms of the polydisperse and three-phase models are the same. As far as the shear modulus is concerned, the three-phase model permits to obtain an expression which, though cumbersome, is valid in any interval of filler concentration. For small volume fractions, this expression also turns into Eq. (II.1), thereby raising certain hopes that this model may be valid in the range of medium concentrations.

$$A\left(\frac{G_c}{G_m}\right)^2 + 2B\left(\frac{G_c}{G_m}\right) + C = 0 \qquad \text{(II.7)}$$

where

$$A = 8\left(\frac{G_{fl}}{G_m} - 1\right)(4 - 5\mu_m)\eta_1 \cdot v_{fl}^{10/3} - 2\left[63\left(\frac{G_{fl}}{G_m} - 1\right)\cdot\eta_2 + 2\eta_1\cdot\eta_3\right]v_{fl}^{7/3}$$

$$+ 252\left(\frac{G_{fl}}{G_m} - 1\right)\eta_2\cdot v_{fl}^{5/3} - 25\left(\frac{G_{fl}}{G_m} - 1\right)(7 - 12\mu_m + 8\mu_m^2)\,\eta_2\cdot v_{fl}$$

$$+ 4(7 - 10\mu_m)\cdot\eta_2\eta_3$$

$$B = -2\left(\frac{G_{fl}}{G_m} - 1\right)(1 - 5\mu_m)\eta_1 v_{fl}^{10/3} + 2\left[63\left(\frac{G_{fl}}{G_m} - 1\right)\eta_2 + 2\eta_1\eta_3\right]v_{fl}^{7/3}$$

$$- 252\left(\frac{G_{fl}}{G_m} - 1\right)\eta_2\cdot v_{fl}^{5/3} + 75\left(\frac{G_{fl}}{G_m} - 1\right)(3 - \mu_m)\,\eta_2\mu_m\cdot v_{fl}$$

$$+ \frac{3}{2}(15\mu_m - 7)\cdot\eta_2\eta_3$$

$$C = 4\left(\frac{G_{fl}}{G_m} - 1\right)(5\mu_m - 7)\,\eta_1 v_{fl}^{10/3} - 2\left[63\left(\frac{G_{fl}}{G_m} - 1\right)\eta_2 + 2\eta_1\eta_3\right]v_{fl}^{7/3}$$

$$- 252\left(\frac{G_{fl}}{G_m} - 1\right)\eta_2 v_{fl}^{5/3} + 25\left(\frac{G_{fl}}{G_m} - 1\right)(\mu_m^2 - 7)\,\eta_2 v_{fl} - 4(7 + 5\mu_m)\,\eta_2\eta_3$$

$$\eta_1 = (49 - 50\mu_{fl}\mu_m)\left(\frac{G_{fl}}{G_m} - 1\right) + 35\,\frac{G_{fl}}{G_m}(\mu_{fl} - 2\mu_m) + 35(2\mu_{fl} - \mu_m)$$

$$\eta_2 = 5\mu_{fl}\left(\frac{G_{fl}}{G_m} - 8\right) + 7\left(\frac{G_{fl}}{G_m} + 4\right)$$

$$\eta_3 = \frac{G_{fl}}{G_m}(8\text{–}10\ \mu\text{m}) + (7\text{–}5\ \mu\text{m})$$

We shall now discuss the use of the regular models in the calculations of the elastic properties of composites [10,11].

Manevich et al. [11] have examined a model for a composite in which the filler particles are of rectangular parallelepipedic shape and are positioned in periodic-lattice sites, with the filler edges parallel to the lattice axes. A similar model is used in [10], but more rigorous assumptions made in the model (absolute rigidity and cubic shape of filler, the matrix positioned only between the parallel faces of the filler) have much reduced its applicability scope. Therefore, the latter model will not be discussed here. The features of the model proposed by Manevich make it possible in principle to analyze the entire range of filling from 0% to 100%, although it may, give rise to errors at high fillings. The model also permits to vary the filler shape from disperse particles to fibers and laminas and, finally, to obtain the averaged characteristics of a composite and the local strain and stress fields (overstress coefficients etc.). The proposed model assumes an orthotropic material. To obtain the characteristics of an isotropic material, an averaging according to the polycrystal analysis principle has to be made. It is shown [11] that different averaging methods (elastic modulus tensor, yielding tensor, self-adjusting method [12]) applied to dispersely filled materials (cubic shape of filler) give similar results. A comparison of the results obtained in terms of such regular model with those of the stochastic (polydisperse and three-phase) models shows that the calculated bulk and shear moduli, and hence all elastic properties of the materials, are much alike. Moreover, experimental elastic moduli of polyethylene filled with glass spheres or alumina are in good agreement with the calculated values, although the studied range of filler concentrations was fairly narrow (from 0 to 24 vol %).

We want to emphasize the results of calculating the maximum stress concentration which is largely model-dependent. The value of this parameter is of interest when analyzing composite failure and, therefore, deserves special attention. However, it should be borne in mind that the proposed solution is for the elastic problem only, and that in the case of heavy deformations (though long before destruction) the elasto-plastic problem needs to be solved.

In the regular model with a spherical filler of uniform size, the stress concentration (in the elastic problem) increases monotonously with the filler bulk content [3,4,13]. In the case of a cube-shaped regularly arranged filler, the stress concentration factor increases only until the filling reaches 7–10 vol %; thereafter it decreases and approaches 1 at 100% filling [11]. This fact implies that composites with thin, but thickness-uniform matrix interlayers are advantageous. In this case the matrix is never overstressed in use, so higher strengths seem to be attainable. Experimental work [14] may be used to illustrate the correctness of such an idea. A composite material consisting of stiff poly(vinyl chloride) with thin interlayers of ethylene-vinyl acetate copolymer was prepared. The elastic modulus of the composite material was at the same level as that of the stiff poly(vinyl chloride) (filler), but the impact strength and ultimate elongation (failure characteristics) were considerably higher than in a conventional mechanical mixture of the same composition. It is interesting to note that the morphological structure of this material has a strong resemblance to the structure of the rubber phase contained in impact-resistant polystyrene and in ABC plastic [15].

Theoretical works examining plastic matrices with disperse fillers are practically nonexistent. Only in one recent paper [16] a cycle of such studies was started. The stress-strain diagram of a model system is analyzed. The model consists of a matrix filled with a regularly arranged, absolutely stiff filler of regular cubic shape, with ideal adhesion. The matrix is assumed to be elastoplastic with a given stress-strain diagram. Numerical calculation of the composite stress-strain diagrams was performed for two filling contents (10 and 30 vol %). A comparison with experimental σ–ε curves has shown a fairly good agreement, thereby giving rise to hope for further

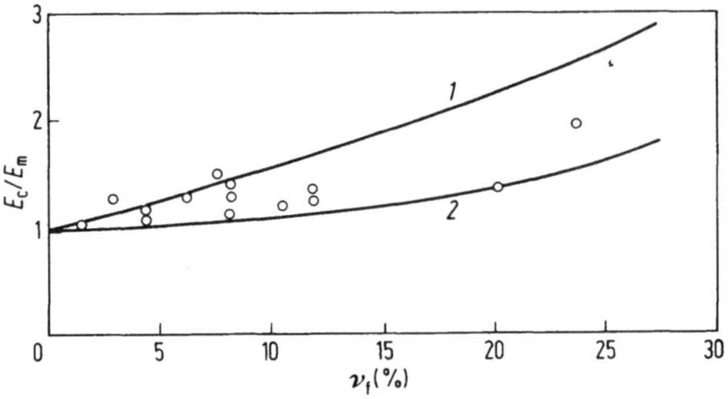

Fig. II.1. Theoretical dependence of the elastic modulus of a model composite on the degree of filling. *1*: calculation for ideal adhesion between matrix and filler; *2*: calculation for complete absence of adhesion; ○: experimental data for polyethylene filled with glassy spheres and alumina [17].

success. Preliminary results demonstrate that the filler (up to 30 vol %) affects but little the plasticity limit which is close to that of the matrix. The proposed calculation technique is expected to allow in future work for finite stiffness, incomplete adhesion, etc.

The effect of the adhesion on the modulus of elasticity of dispersely-filled composites has also been studied theoretically [17,18]. Numerical calculations were carried out for the regular model (elastic matrix with a regularly arranged, stiff cubic-shaped filler). The dependence of the elastic modulus of the composite on the degree of filling was calculated for ideal adhesion and for complete absence of adhesion (see Fig. II.1). A significant decrease of the elastic modulus is seen in the absence of adhesion. Figure II.1 shows also the experimental result for polyethylene filled with non-finished glass spheres and alumina. The experimental data are located between the two theoretical curves.

A stochastic model with definite limitations on particle geometry and arrangement in space was proposed by Gai, Manevich and Oshmyan [19]. Thus far, the problem has been solved only for a two-dimensional plane (a three-dimensional version of the

Fig. II.2. Diagram of a plane model of a composite with randomly arranged elements of matrix (light squares) and filler (dark squares) [19].

problem is being analyzed at present). A linear elasticity theory for an inhomogeneous medium in the form of a rectangle (Fig. II.2), consisting of randomly arranged elements of soft matrix and stiff filler was developed. A set of elastic theory equations was integrated assuming equality of stress and displacement, i.e. ideal adhesion conditions (this condition is obligatory for all problems of stochastic models). The problem was solved numerically by the finite-element method, and the results obtained were analyzed in terms of the percolation theory, which is usually applied to analyze electrical and thermal conductivities and mass transfer in such heterogeneous systems. The analysis has shown that a correlation between the elastic properties of a composite and the geometric percolation characteristics is observed only if the elastic modulus of the matrix differs substantially from that of the filler. Thus, a sharp increase of the composite elastic modulus and a minimum of Poisson's ratio are observed at 40 vol % of filling (plane model, see Fig. II.1), i.e. at the point where the continuous phase of soft matrix elements disappears. (It should be noted that the model assumes absolutely stable bonds between the filler particles in any filler cluster.) At this point the calculations resulted in a μ_c value lower than μ_{fl} and μ_m. This means that the filler clusters are resistant to transverse compression due to uniaxial tension of the composite, while the matrix usually suffers a considerable three-axial tension.

It is probably too early to speak of the applicability of the model; it needs further theoretical analysis and comparison with thoroughly designed experiments.

When analyzing the problem of solid particle suspension viscosity (cf. Eq. II.4), it should be borne in mind that, although some similiarity with the problem of solid body linear elasticity is observable, the analogy is far from being complete. For example, non-Newtonian effects are very noticeable in the case of the flow of suspensions of medium and high concentrations, even if the liquid proper is purely Newtonian [20]. Therefore, extrapolation to zero shear rate has to be made when analyzing experimental data. Certainly, these effects are important for polymer matrices showing great deviations from the Newtonian flow laws.

Moreover, at small deformations of a composite the filler particle arrangement varies only insignificantly, whereas in a suspension flow the particle arrangement is determined by the flow conditions. Therefore no analogy has been observed in this case.

McGee and McCullogh [21] have compared various theoretical and semiempirical models for calculating the elastic modulus of particulate-filled composites with each other and with experimental data [22–25,5,9]. It turned out that, up a filling of 20 to 30 vol %, most models agreed with each other and could correctly describe the experimental data (see Fig. II.3). The differences between models become apparent at higher filler concentrations (40–50 vol %). Three-phase and polydisperse models describe the experimental data quite well. But their main disadvantage, the assumption of a specially broad filler size distribution enabling a 100 % space filling, becomes effective at filler concentrations > 50 %. The experiments, carried out with much narrower filler size distributions have led to higher values of Young's modulus compared with the values calculated in terms of these models.

Filler particle interactions were takes into account in many semiempirical works (Halpin-Tsai, Nielsen, Kerner et al.) by introducing a correction to the filler bulk fraction which satisfies two limiting conditions:

$$\tilde{v}_{fl} \rightarrow v_{fl} \quad \text{at} \quad v_{fl} \rightarrow 0; \quad \text{and} \quad \tilde{v}_{fl} \rightarrow 1 \quad \text{at} \quad v_{fl} \rightarrow v_{fl}^*$$

where

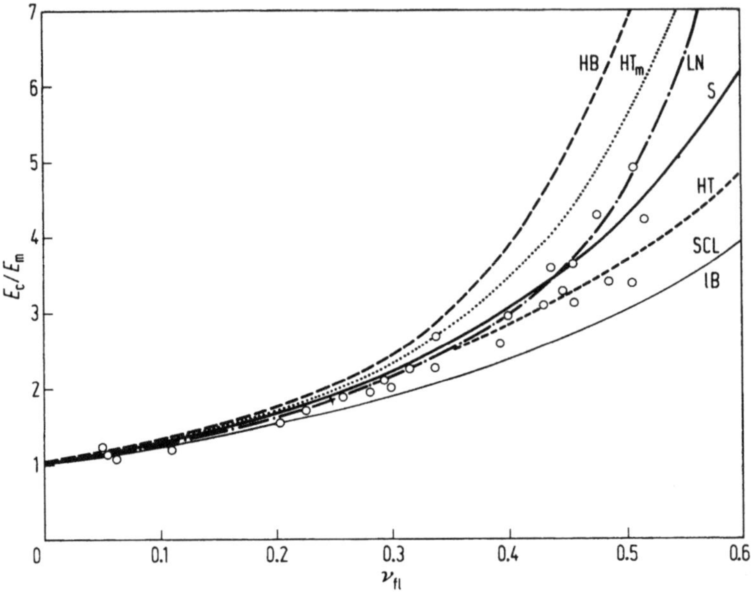

Fig. II.3. Comparison of experimental data of Ishai and Cohen [32] (open circles) with the value of E_c predicted by selected models; (HB): singly embedded model, Hill [22] and Budianskii [23]; (HT_m): Halpin-Tsai Equation with Lewis-Nielsen Equation [26]; (S): S-combining Rule [21]; (HT): Halpin-Tsai Equation [24,25]; (SCL): three-phase model of Smith [8] and Christensen and Lo [9]; (1B) is improved lower bound [21].

v_{fl} — is the true bulk fraction of filler, $\tilde{v}_{fl}$ — is the corrected bulk fraction of filler, v^*_{fl} — is the maximum admissible filling bulk fraction for a given filler.
In particular, the following expression for $\tilde{v}_{fl}$ was proposed [26,27]:

$$\tilde{v}_{fl}/v_{fl} = 1 + v_{fl}(1 - v^*_{fl})/v^{*2}_{fl}$$

The authors of reference [21] proposed another way of calculating the composite elastic properties. Their main idea was to present a composite either as a polymer which is a continuous phase with a filler in the form of inclusions at $v_{fl} < v^*_{fl}$, or as a filler which is a continuous phase with a polymer in the form of inclusions, at $v_{fl} > v^*_{fl}$. Two limiting cases, a stiff filler in a soft matrix and a soft filler in a stiff matrix were examined.

The limiting expressions which may prove to be most important in practice will be given here. For foams (soft fill in hard matrix) Young's modulus is given by:

$$E_c = v_m^2 E_m (1 - v_{fl}\Phi_{Em}) \tag{II.8}$$

where

$$\Phi_{Em} = \frac{1}{4}\left\{1 + 2\mu_m\left[\frac{2v^*_{fl} - 1}{v_{fl}}\right]\right\}$$

($v_{fl}^* = 0.74$ for hexagonal densest packing of spheres, $v_{fl}^* = 2/3$ for a statistically dense random packing of spheres).

For $v_{fl} < 0.5$ expression (II.8) may be approximated by the formula $E_c \approx E_m \cdot v_m^2$, which was experimentally well confirmed in [28].

For polymers with mineral fillers (stiff filler in soft matrix) another limiting conditions, $E_{fl} \gg E_m$ is satisfied. Then, the following expressions are obtained for bulk modulus K_c and shear modulus G_c:

$$\frac{K_c}{K_m} = \frac{1 + v_{fl} \dfrac{K_{fl} - K_m}{K_{fl} + \dfrac{2(1 - 2\mu_m)}{1 + \mu_m} \cdot K_m} \cdot \dfrac{2(1 - 2\mu_m)}{1 + \mu_m}}{1 - v_{fl} \dfrac{K_{fl} - K_m}{K_{fl} + \dfrac{2(1 - 2\mu_m)}{1 + \mu_m} K_m} \cdot \psi} \tag{II.9}$$

$$\frac{G_c}{G_m} = \frac{1 + v_{fl} \dfrac{G_{fl} - G_m}{G_{fl} + \dfrac{7 - 5\mu_m}{8 - 10\mu_m} G_m} \cdot \dfrac{7 - 5\mu_m}{8 - 10\mu_m}}{1 - v_{fl} \dfrac{G_{fl} - G_m}{G_{fl} + \dfrac{7 - 5\mu_m}{8 - 10\mu_m} G_m} \cdot \psi} \tag{II.10}$$

with

$$\psi \approx 1 + \left(\frac{v_m}{v_{fl}^*}\right) (v_{fl} v_{fl}^* + v_m(1 - v_{fl}^*)) \tag{II.11}$$

K_c and G_c having been calculated, Young's modulus E_c and Poisson's ratio μ_c can be readily obtained as:

$$E_c = \frac{9KG}{3K + G} \quad \text{(II.12)} \quad \text{and} \quad \mu = \frac{3K - 2G}{6K + 2G} \tag{II.13}$$

These equations describe quite well the experimental data [29-33] for a filler with a narrow size distribution (glass spheres), up to 50 vol % filling.

2.2 Calculation of the Elastic Properties of Fiber and Lamina Reinforced Composites

In the case of fibrous and laminated composites, the orientation, including the filler particle mutual orientation, needs to be taken into consideration, which makes the theoretical analysis of such systems more difficult.

In the analysis of the linear elasticity of fibrous and laminated composites, two types of models (regular (I) and stochastic (II) are also used. Uniaxial systems are usually analyzed first in either model, whereupon the directions are averaged in a definite way assuming that a composite consists of randomly oriented uniaxial domains. It is difficult to estimate how well this procedure can describe the randomly oriented fiber system, but it seems to be often quite workable when applied to commercial materials.

In this Section we shall discuss the following types of fillers: uniaxial continuous fibers, interwoven continuous fibers, short fibers, and laminas.

2.2.1 Continuous Fiber Reinforced Composites

First, a composite element consisting of a matrix and parallel continuous fibers will be discussed. Such an element is inherently anisotropic and, therefore, its linear elastic behavior should be described not by two constants, as in case of an isotropic material, (E and μ), but by a rigidity or pliability matrix defining the relationship between the strain and stress tensor components.

For a material which is isotropic in a plane perpendicular to the fiber axes, i.e. in case of a random fiber packing in the composite element cross section, we get:

$$\{\varepsilon_i\} = [S_{ij}]\{\sigma_j\} \tag{II.14}$$

$$[S_{ij}] = \begin{bmatrix} \frac{1}{E_{11}} & -\frac{\mu_{21}}{E_{22}} & -\frac{\mu_{21}}{E_{22}} & 0 & 0 & 0 \\ -\frac{\mu_{12}}{E_{11}} & \frac{1}{E_{22}} & -\frac{\mu_{32}}{E_{22}} & 0 & 0 & 0 \\ -\frac{\mu_{12}}{E_{11}} & -\frac{\mu_{23}}{E_{22}} & \frac{1}{E_{22}} & 0 & 0 & 0 \\ 0 & 0 & 0 & \frac{1}{G_{12}} & 0 & 0 \\ 0 & 0 & 0 & 0 & \frac{1}{G_{23}} & 0 \\ 0 & 0 & 0 & 0 & 0 & \frac{1}{G_{12}} \end{bmatrix}$$

where ε_i and σ_i are components of the strain and stress tensors; $[S_{ij}]$ is the pliability matrix; E_{11} and E_{22} are the elastic moduli for longitudinal and transverse tensions; G_{12}, G_{23}, μ_{12}, μ_{21}, μ_{23}, μ_{32} are the respective shear moduli and Poisson ratios; μ_{12} is the shear modulus in the plane of the fibers.

To calculate the properties of such an element of a composite, the geometrical structure and fiber cross section packing should be specified. As in the description of dispersely filled composites, either a plane regular model, a stochastic polydisperse model (see Fig. II.4), or a three-phase model can be used. Since the real fibrous composites consist of fibers with a very narrow diameter distribution (almost all fibers are of the same diameter), regular models are preferable. However, a quantitative analysis for systems with realistic filling ratios ($v_f \simeq 70\%$) and stiffnesses ($E_f/E_m \approx 10$

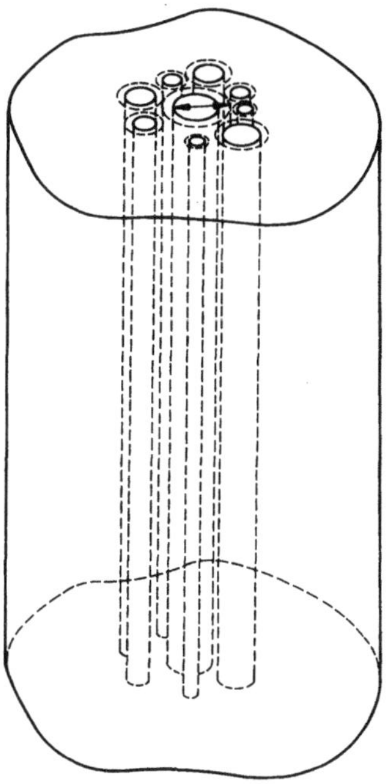

Fig. II.4. Polydisperse model of fiber-reinforced composites.

— 100) has shown that the calculation results obtained in terms of the various models, including the regular model with hexagonal packing, differ only little [34,35].

Accurate calculations for the fiber-aligned elastic modulus E_{11} and for the Poisson ratio, μ_{12} have shown that the following equations are good approximations:

$$E_{11} \simeq v_f E_f + v_m \cdot E_m \qquad \text{(II.15)}$$

$$\mu_{12} \simeq v_f \mu_f + v_m \mu_m \qquad \text{(II.16)}$$

Small deviations from additivity are due to the difference between the Poisson ratios of matrix and filler and to the relevant radial interaction in case of axial tension.

Complete expressions for composite elastic constants were obtained by Hill and Hashin [36,37]; some expressions for other constants may also be found there. In our opinion, the following equation for the shear modulus in a plane of fibers is of particular interest:

$$\frac{G_{12}}{G_m} = \frac{G_f(1 + v_f) + G_m(1 - v_f)}{G_f(1 - v_f) + G_m(1 + v_f)} \qquad \text{(II.17)}$$

The problem to calculate the transverse elastic modulus E_{22} is qualitatively similar to the determination of the dispersely-filled composite elastic modulus. The above-mentioned results (Section 2.1) may be used as a rough estimate.

Numerical calculations and analytical estimates of the properties of fibrous composites for regular models are presented in [38–40]. Such calculations make it possible to obtain the strain and stress fields, and hence the mean macroscopic characteristics of a composite, and the local microscopic parameters (the maximum stress concentrations, etc.). These parameter most probably define the processes occurring in case of relatively heavy deformations, i.e. plasticity and failures.

2.2.2 Interwoven Fiber Reinforced Composites

In the case of reinforcement by interwoven continuous fibers (fabrics), the fibers are periodically bent and, accordingly, the elastic modulus along the reinforcement directions, E_{11}, decreases. Certainly, other elastic constants of the material are also changed. We present here only the estimate of the change of the elastic modulus, E_{11} in periodically bent fibres according to [41]:

$$\frac{E_{11}^{\varphi}}{E_{11}^{0}} = 1 - 2\left[1 - 2\,\frac{G_m}{E_f}\cdot\frac{1}{v_f(1 - v_f)}\right]\cdot\varphi^2 \qquad \text{(II.18)}$$

where φ is the angle of distortion. Expression (11.18) is valid for $\varphi < 10°$.

2.2.3 Effect of the Fiber Ends on the Elastic Properties

Now we shall proceed to describing the mechanical properties of a unidirectional element consisting of a matrix and parallel fibers of finite sizes. The calculations for this problem are usually made by the so-called shear analysis, on the assumption that a load is transmitted from the matrix to the fiber only through tangential shear stresses. Although simplified by fairly rough assumptions, such approach is worth describing. The fiber ends are supposed to be unloaded, and the distribution of tensile stresses along the fibers (for a stretched composite) is of the form shown in Fig. I.1. The tensile stress in each fiber rises gradually when moving from its ends to middle. The law, according to which the load rises, depends on the particular model (elastic or elastoplastic matrix, effect of neighboring fibers, etc.). Such an approach leads to the introduction of the concept of the "critical length" of the reinforcing fiber (cf. Section 1.1.1). A part of the fiber is insufficiently loaded, thereby reducing the reinforcing effect of the filler. The aim of the various models is to estimate the critical length of the fiber.

For example, in the Kelly-Tyson model [42] there are plasticity regions in the matrix around the fiber ends, due to high tangential loads. Tangential stresses transferring the load to the fibers are defined by the matrix yield point and remain constant over the entire plasticity region at the fiber ends. Moreover, it is assumed that the plasticity region is sufficiently large and that in practice all the load is transferred to the fiber in this region. From the condition of equilibrium in any cross section of a fiber at a distance from its end, we obtain the load in the fiber: $P_f = \pi d_f \tau_m$ if the stress is $\sigma_f = 4\tau_m \cdot x/d_f$. The stress in the fiber is unambiguously related to its strain: $\sigma_f = E_f \varepsilon_f$; it rises until the strains in the matrix and in the fiber are equal. The plasticity region comes to an end at this point, whereupon the fiber stress remains constant (see Fig. I.1). By definition, this point corresponds to $x = l_c/2$ (l_c is the critical length of the fiber). We find:

$$l_c = d_f \frac{\sigma_f}{2\tau_m} = d_f \frac{E_f \varepsilon_f}{2\tau_m} = d_f \frac{E_f \varepsilon_m}{2\tau_m} \qquad \text{(II.19)}$$

This equation is analogous to Eq. (I.8) in Section 1.1.1., but here it is given for small ε values.

The mean stress $\bar{\sigma}_f$ in a fiber, taking into account the linear drop of the stress in the zones $\times$ at its ends, is given by:

$$\bar{\sigma}_f = \frac{1}{l}\int_0^l \sigma_f \, dx = \sigma_f \left(1 - \frac{l_c}{2l}\right) = E_f \varepsilon_m \left(1 - \frac{l_c}{2l}\right) \quad \text{at} \quad l > l_c \qquad \text{(II.20)}$$

$$\bar{\sigma}_f = \frac{\tau}{d_f} \cdot l = E_f \varepsilon_m \frac{l}{2l_c} \quad \text{at} \quad l < l_c \qquad \text{(II.21)}$$

Further, we can calculate the elastic modulus of a unidirectional composite containing v_f fibers of length l. It should be noted that in this case the stress is an additive value:

$$\sigma_c = \bar{\sigma}_f v_f + \sigma_m v_m = E_c \varepsilon_c \approx E_c \varepsilon_m \qquad \text{(II.22)}$$

whence we obtain

$$E_{11} = E_m v_m + E_f v_f \left(1 - \frac{l_c}{2l}\right) \quad \text{at} \quad l > l_c \qquad \text{(II.23)}$$

$$E_{11} = E_m v_m + E_f v_f \frac{l}{2l_c} \quad \text{at} \quad l < l_c \qquad \text{(II.24)}$$

Other models has been used for the calculation of l_c based on shear analysis [43-46]. Rosen [45] gave the following expression for the fiber critical length, in the case of an elastic matrix:

$$\frac{l_c}{d_f} \approx 1{,}15 \left[\frac{1 - v_f^{1/2}}{v_f^{1/2}} \left(\frac{E_f}{G_m}\right)^{1/2}\right] \qquad \text{(II.25)}$$

All these models are fairly rough but permit a qualitative analysis of the effect of the properties of matrix and fibers on the reinforcing action of the filler. It should, however, be noted that for very small deformations, the models based on elastic analysis (Eq. II.25) are more appropriate than the Kelly-Tyson model, because in this case large regions of plasticity do not arise on the fiber ends. In fact, inspection of Eq. (II.19) shows that in the Kelly-Tyson plastic model the fiber critical length l_c depends on the deformation quantity ε. As the strain level decreases, l_c decreases and the reinforcing effect increases. The solution of the linear-elastic problem (see Eq. II.25) implies that the fiber critical length is constant and independent of the composite strain level. Thus, as the strain level decreases, the plasticity region also decreases and, eventually, the main load is transferred to the fibers through the elastic region of tangential

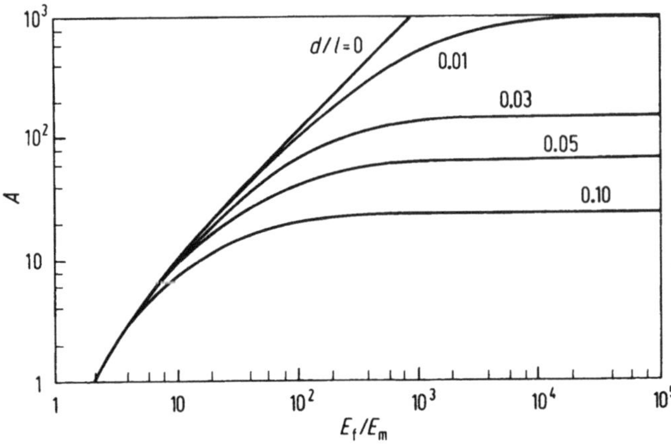

Fig. II.5. Influence of fiber length on the reinforcement efficiency factor A $\mu_f = 0.25$; $\mu_m = 0.35$ [51].

stresses. If, however, the concept of critical length is used to examine the failure process, it is assumed that the maximum reinforcing effect is observed when fibers just begin to break and, therefore, the expression $\sigma_f = E_f \varepsilon_f \simeq E_f \varepsilon_m$ in Eq. (II.19) can be replaced by the ultimate fiber strength σ_f^*. The resulting equation $l_c = d_f \sigma_f^* / 2\tau_m$ has a more fundamental meaning, since considerable plasticity zones will have developed on the fiber ends, at this point, so that models considering these zones appear to be justified.

A vigorous solution for the problem of the elasticity modulus of a discrete fiber reinforced composite is given by Russel [47]. However, although derived only for small bulk fractions of fillers, the expressions are very complicated. We shall only present graphically the reinforcement efficiency factor A in the expression

$$\frac{E_{11}}{E_m} = 1 + v_f \cdot A \qquad \text{(II.26)}$$

plotted against the filler-to-matrix elastic modulus ratio, at various diameter-to-length ratios (see Fig. II.5).

All models described above neglect the stressed state of the fiber ends. Hence they should be used only for relatively long fibers where such stresses are insignificant. Besides, in case of heavy deformations approaching failure the fiber strand ends are probably of decisive importance.

2.2.4 Transition from Unidirectional to Randomly Reinforced Composites

Having obtained expressions for the parameters of a unidirectional element of a composite, we may now proceed to the parameters of an isotropic material consisting of randomly oriented unidirectional elements i.e. to the process of averaging.

Averaging problems have been solved more or less rigorously by various authors [48–52]. The most rigorous approach seems to have been described by Christensen [51]. Exact formulae prove usually [52] to be very complicated and inconvenient for practical use. We shall present here only those approximate expressions which are of the greatest practical importance.

The modulus of a composite is given by:

$$E_c \approx \frac{1}{6} v_f E_f + v_m E_m \tag{II.27}$$

for randomly oriented continuous fibers and

$$E_c \approx \frac{1}{3} v_f E_f + v_m E_m \tag{II.28}$$

for random orientations in a plane. (The second term in these equation is introduced in order to preserve their validity when the filler bulk fraction decreases to zero.) The approximate expressions II.27 and II.28 are insensitive to the values of the Poisson ratio. Certainly, they are valid for sufficiently stiff fibers ($E_f \gg E_m$).

Neglecting matrix contributions, the following approximate equations were obtained for shear modulus and Poisson's ratio for fibers with random orientation in a plane:

$$G_c \approx \frac{1}{8} v_f \cdot E_f + v_m G_m \tag{II.29}$$

$$\mu_c \approx \frac{1}{3} \tag{II.30}$$

For fibers with random bulk orientations the corresponding equation is

$$G_c \approx \frac{1}{16} v_f E_f + v_m G_m \tag{II.29'}$$

(The second term, $v_m G_m$, is introduced in Eqs. (II.29) and (II.29′) to provide for correct extrapolation to zero filling.)

2.2.5 Lamina-reinforced Composites

The practical application of such systems is confined to mica. Theoretically, such composites are advantageous compared with fibrous materials in that they have a higher reinforcing effect at the same filler bulk fraction. They also show isotropy of in-plane properties, and a considerably lower permeability for gases and liquids.

Therefore, a further development of this type of composites and an expansion of their application scopes may be expected.

In the case of planar reinforcement, the following equations are approximately satisfied for the elastic modulus in the plane of reinforcement, and for the shear modulus in a plane perpendicular to the reinforcement plane, for the pliability (the inverse elastic modulus) perpendicular to the reinforcement plane, and for the inverse shear modulus in the reinforcement plane [53]:

$$E_c'' \approx v_m E_m + v_f E_f \tag{II.31}$$

$$G_c^{\perp} \approx v_m G_m + v_f G_f \tag{II.32}$$

$$\frac{1}{E_c^{\perp}} \approx \frac{v_m}{E_m} + \frac{v_f}{E_f} \tag{II.33}$$

$$\frac{1}{G_c''} \simeq \frac{v_m}{G_m} + \frac{v_f}{G_f} \tag{II.34}$$

These relations are rigorously satisfied if the Poisson ratios of the components are the same, while increasing deviations occur with increasing difference between the Poisson ratios of matrix and filler.

An interesting result has been obtained for an isotropic medium with laminas randomly oriented in bulk and with averaged elastic properties [53]. Independently of the values of the Poisson ratios, the following approximate equation for the elastic modulus proved to be applicable:

$$E_c \simeq \frac{1}{2} v_f E_f + v_m E_m$$

Compared with the expression for a fibrous filler, this equation indicates a threefold increase of the reinforcing effect when a lamina filler is used. The same is certainly valid for the shear modulus of such a material. More rigorous expressions for the elastic characteristics of composites with lamina fillers are given in ref.[51] Shear analysis was used to determine the effect of the size (characteristic ratio) of laminas [54]. The same approach used by Cox [42] to study the elastic modulus of short-fiber composites was used by Padawer and Beecher [54] to analyze the elastic modulus of flake filler reinforced composites. They obtained the following expression for a factor A correcting for the decrease of the elastic modulus of a transversely isotropic plane composite due to a limited size of flakes:

$$A = 1 - \frac{\tan hU}{U} \tag{II.35}$$

where $U = \alpha \left[\frac{G_m v_f}{E_f v_m}\right]^{1/2}$, and α is the diameter-to-thickness ratio (the characteristic ratio) of the flakes. The flakes were assumed to be uniformly arranged in the composite. In case of random overlapping of the flakes, Riley [55] has obtained the expression

$$A = 1 - \frac{\ln (U + 1)}{U} \tag{II.36}$$

Thus, the expression

$$E_c'' \approx v_f E_f \left[1 - \frac{\ln (U + 1)}{U}\right] + v_m E_m \tag{II.37}$$

is valid for the elastic modulus of a composite with small flakes aligned in the composite reinforcement plane. For the elastic modulus of an isotropic composite with randomly oriented small flakes the expression becomes:

$$E_c \approx \frac{1}{2} v_f E_f \left[1 - \frac{\ln (U + 1)}{U}\right] + v_m E_m \tag{II.38}$$

In the case of lamina-filled composites with a plastic matrix, an approach analogous to that of Kelly-Tyson has been made [56]. The following expression was obtained for the elastic modulus in the reinforcement plane of such a composite:

$$E_c'' \approx 0{,}56 \cdot v_f E_f \left(1 - \frac{E_f \varepsilon_m}{6\alpha \tau_m}\right) + v_m E_m \tag{II.39}$$

The factor 0.56 takes care of random overlapping of the filler flakes. As in the Kelly-Tyson model, it should be noted that the matrix plasticity becomes noticeable only when the general deformation level rises, so expressions of this type should be applied to heavy deformations and to ultimate failure stresses. It should be borne in mind, however, that the matrix plasticity near a filler particle may become noticeable even if the strain is not very high. Therefore the pure elastic analysis may prove to be valid only at very small (<0.1%) average strains.

2.2.6 Short Fiber Reinforced Composites

The exact solutions for the elastic properties of composites with randomly oriented short fibers (the most interesting case in practice) were presented by Christensen [51]. To obtain numerical results for particular systems, numerous tedious calculations have to be made, because any real system shows a broad fiber length distribution, a complicated nonuniform bulk distribution of the fibers, and a complicated orientation of the fibers. It seems to us that it is quite reasonable to use approximate formulae of estimative nature.

In particular, in the elastic modulus expressions, reduced values of $E_f\left(1 - \frac{l_c}{2l}\right)$ at $l > l_c$ and of $E_f \frac{l}{2l_c}$ at $l < l_c$ should be substitute for E_f (cf. Eqs. II.20 and II.21).

Thus, we obtain for randomly oriented short fiber reinforcement:

$$E_c \approx \frac{1}{6} v_f E_f \left(1 - \frac{l_c}{2l}\right) + v_m E_m \quad \text{at} \quad l > l_c \tag{II.40}$$

$$E_c \approx \frac{1}{6} v_f E_f \frac{l}{2l_c} + v_m E_m \qquad \text{at} \quad l < l_c \tag{II.41}$$

$$G_c \approx \frac{1}{16} v_f E_f \left(1 - \frac{l_c}{2l}\right) + v_m E_m \quad \text{at} \quad l < l_c \tag{II.42}$$

$$G_c \approx \frac{1}{16} v_f E_f \frac{l}{2l_c} + v_m E_m \qquad \text{at} \quad l < l_c \tag{II.43}$$

In the case of short fiber reinforcement in a plane, the corresponding equations are:

$$\left.\begin{aligned} E_c &\approx \frac{1}{3} v_f E_f \left(1 - \frac{l_c}{2l}\right) + v_m E_m \\ G_c &\approx \frac{1}{8} v_f E_f \left(1 - \frac{l_c}{2l}\right) + v_m G_m \end{aligned}\right\} \quad \text{at} \quad l > l_c \tag{II.44}$$

$$\left.\begin{aligned} E_c &\approx \frac{1}{3} v_f E_f \frac{l}{2l_c} + v_m E_m \\ G_c &\approx \frac{1}{8} v_f E_f \frac{l}{2l_c} + v_m G_m \end{aligned}\right\} \quad \text{at} \quad l < l_c \tag{II.45}$$

In order to make allowance for the fiber length distribution function, the relevant expressions have to be integrated, for example:

$$E_c \approx \frac{1}{6} v_f E_f \left[\int_0^{l_c} \frac{l}{2l_c} Q(l)\, dl + \int_{l_c}^{\infty} \left(1 - \frac{l_c}{2l}\right) Q(l)\, dl \right] + v_m E_m \tag{II.46}$$

where $Q(l)$ is a normalized function of fiber length distribution, i.e. $\int_0^{\infty} Q(l)\, dl = 1$.

Injection molded material is anisotropic with an axial symmetry. Its properties may be estimated by averaging the fibre orientation distribution function by, for example, the method described in [51].

2.3 Calculation of Thermal and Electric Conductivities and mass Transfer Processes in Polymer Composites

The equations for thermal and electric conductivity, as well as for gas and moisture diffusion in composites are identical. Moreover, all these processes involve at least two different limiting types of events, with two different approaches to solving the relevant problems. When analyzing, for example, the electric conductivity, "contact" conduction should be emphasized, if the carriers of the current are concentrated in the filler particles and their macroscopic motion is only possible via direct contacts of the particles, giving rise to the transfer of a carrier of current from one filler particle to another. "Normal" conductivity, where the carriers of current travel in each of the phases and may traverse the boundary between the two phases, is the other extreme case.

Usually, the "contact" conductivity is theoretically analyzed using the percolation method, which deals with the probabilities of clustering of contiguous particles. In this case the theory predicts a characteristic critical filler concentration (percolation limit) at which the conductivity shows an abrupt change. The contact mechanism of conductivity implies a key importance of the spatial arrangement of the filler particles. The geometric ordering of certain fillers may be mentioned in this connection. For example, soot particles in polymer matrices are structured due to their specific surface properties and to the adsorption-active regions on their surfaces. Cross-linked filled systems may be prepared by low-temperature sintering of polymer particles whose surfaces are covered by much smaller metal particles. In general, geometric ordering should be allowed for when developing a conductivity theory for such systems.

The "normal" conductivity is analyzed using the regular and stochastic models mentioned in Section 2.1 when examining the linear elasticity of composites.

Hashin and Shtrikman [60] solved the problem of "normal" thermal conductivity of a composite with spheric inclusions in terms of the polydisperse model and obtained the following expression for the thermal conductivity coefficient:

$$\varkappa_c = \varkappa_m \left[1 + \frac{v_{fl}}{\frac{1}{3}(1 - v_{fl}) + \frac{\varkappa_m}{\varkappa_{fl} - \varkappa_m}} \right] \quad \text{(II.47)}$$

The same expression was obtained earlier by Kerner [51] for the electric conductivity, and also by Christensen [51] who used the three-phase model, thereby indicating again that the polydisperse and the three-phase models are equivalent.

If a polymer matrix is filled with randomly oriented short fibers or laminas, the properties of the material are direction-independent, i.e. the composite is isotropic. In theoretical treatments such materials are usually presented as a combination of randomly oriented domains of a unidirectional anisotropic composite. Therefore, the parameters of such a unidirectional structural element must first be calculated.

The fiber aligned conductivity of such an element is an additive value; the thermal conductivity coefficient, for example, is given by:

$$\varkappa_c'' = \varkappa_{fl} v_{fl} + \varkappa_m v_m \quad \text{(II.48)}$$

To calculate the transverse conductivity one has to use a certain structural geometric model of fiber arrangement in a plane perpendicular to the fiber axis. In this case also regular and stochastic models are used. The polydisperse [62] and three-phase [51] models both yield the following expression for the transverse thermal conductivity coefficient:

$$\varkappa_c^{\perp} = \varkappa_m \left[1 + \frac{v_{fl}}{(1 - v_{fl})/2 + \varkappa_m/(\varkappa_{fl} - \varkappa_m)} \right] \quad \text{(II.49)}$$

In the case of a composite element consisting of parallel layers of matrix and filler, the two-directional conductivity in the reinforcement plane is given by eq. (II.48), whereas in the direction perpendicular to the reinforcement plane, resistivity, rather than conductivity, is an additive value, i.e.:

$$\frac{1}{\varkappa_c^{\perp}} = \frac{v_{fl}}{\varkappa_{fl}} + \frac{v_m}{\varkappa_m} \tag{II.50}$$

Knowing the conductivity parameters for the unidirectional elements of structure, one can calculate those of the isotropic material by averaging over all directions.

Let a unidirectional element be cross-symmetrical and characterized by the two conductivities $\varkappa''$ and $\varkappa^{\perp}$. Then, the upper and lower limits of conductivity of the isotropic composite are:

$$\frac{\varkappa^{\parallel}\varkappa^{\perp}}{\frac{1}{3}\varkappa^{\parallel} + \frac{2}{3}\varkappa^{\perp}} < \varkappa_c < \frac{1}{3}\varkappa^{\parallel} + \frac{2}{3}\varkappa^{\perp} \tag{II.51}$$

The calculation of the properties of composites in terms of the regular model may be exemplified by the results obtained by Brydges et al. [63] who calculated the factors of gas and fluid penetration through a composite based on a polymer matrix reinforced with parallel glass bands. The glass bands were assumed to be impermeable. The following expression was obtained for the permeability coefficient of the composite:

$$\frac{\chi_c}{\chi_m} = \frac{1}{1 + \frac{b}{2g} + \left(\frac{b}{a}\right)^2 \left(\frac{v_{fl}^2}{1 - v_{fl}}\right) \alpha(1 - \alpha)} \tag{II.52}$$

where $2g$ is the distance between glass bands.

The equation shows that in this case a form factor (the glass band width-to-thickness ratio b/a) is important. Reinforcement with bands with a reasonable value of the form factor ($b/a \approx 10 \div 100$) reduces the permeability of the composites by $2 \div 3$ orders of magnitude compared with that of fibrous composites (for which $b/a \simeq 1$).

2.4 Calculation of the Thermal Expansion Coefficient of Composites

Another problem of great practical interest is the calculation of the thermal expansion coefficient [64]. For macroscopically isotropic composites (composites filled either with particles or with randomly oriented fibers or laminas), the following expression has been obtained for the thermal expansion coefficient:

$$\alpha_c = \bar{\alpha} + \frac{\alpha_{fl} - \alpha_m}{\frac{1}{K_{fl}} - \frac{1}{K_m}} \left[\frac{1}{K_c} - \overline{\left(\frac{1}{K}\right)}\right] \tag{II.53}$$

where

$$\bar{\alpha} = v_{fl}\alpha_{fl} + v_m\alpha_m; \qquad \overline{\left(\frac{1}{K}\right)} = \frac{v_{fl}}{K_{fl}} + \frac{v_m}{K_m}$$

K_{fl}, K_m, and K_c are the bulk moduli of filler, matrix and composite, respectively. The value of K_c may be determined experimentally or calculated.

A transversely anisotropic medium (a fibrous composite with parallel fibers distributed randomly in a cross section; a composite randomly reinforced with fibers in a plane; a composite reinforced with laminas in a plane) expands anisotropically when heated. It is characterized by two coefficients of thermal expansion along and perpendicular to the axis, $\alpha_c^{\parallel}$, and $\alpha_c^{\perp}$ [51]:

$$\alpha_c^{\parallel} = \bar{\alpha} + \frac{\alpha_f - \alpha_m}{\dfrac{1}{K_f} - \dfrac{1}{K_m}} \left[\frac{3(1 - 2\mu_{12})}{E_{\parallel}} - \overline{\left(\frac{1}{K}\right)}\right] \tag{II.54}$$

$$\alpha_c^{\perp} = \bar{\alpha} + \frac{\alpha_f - \alpha_m}{\dfrac{1}{K_f} - \dfrac{1}{K_m}} \left[\frac{3}{2K_{23}} - \frac{3\mu_{12}(1 - 2\mu_{12})}{E_{\parallel}} - \overline{\left(\frac{1}{K}\right)}\right] \tag{II.55}$$

where $\bar{\alpha}$ and $\overline{\left(\frac{1}{K}\right)}$ have the same meaning as in (II.53); E_{11}, K_{23}, and μ_{12} are elastic characteristics of the composite, namely the elastic modulus for stretching along the axis, the bulk modulus in the case of planar deformation, and Poisson's ratio in the case of stretching along the axis, respectively. The elastic characteristics may be determined either experimentally or theoretically (see Section 2.2).

It should be noted that the problem of composite swelling in a certain medium (e.g. water) is formally the same as the problem of thermal expansion described above. However, swelling at the interface may present a different problem, in particular if composites based on hydrophobic polymers and solid mineral fillers are swollen in water.

The equations for the thermal expansion coefficients of unidirectional fiber-reinforced composites were obtained by Greszczuk [65] in the following form:

$$\alpha_c'' = \frac{1}{E_c''} [\alpha_m v_m \cdot E_m + \alpha_f v_f E_f] \tag{II.56}$$

$$\alpha_c^{\perp} = \frac{1}{E_c^{\perp}} \left[\alpha_0 E_0 \sqrt{\frac{v_f}{\pi}} + \alpha_m E_m \left(1 - \sqrt{\frac{v_f}{\pi}}\right)\right] \tag{II.57}$$

where

$$\alpha_0 = \alpha_m\left(1 - 2\sqrt{\frac{v_f}{\pi}}\right) + 2\alpha_f\sqrt{\frac{v_f}{\pi}} - \mu(\alpha_f - \alpha_m)\left(1 - 2\sqrt{\frac{v_f}{\pi}}\right)$$

$$E_0 = \frac{E_m \cdot E_f}{E_f\left(1 - 2\sqrt{\frac{v_f}{\pi}}\right) + 2E_m\sqrt{\frac{v_f}{\pi}}}$$

The Eqs. (II.54) and (II.56), as well as (II.55) and (II.57) give similar results [65].

Concluding this Section, it should be noted that the theories based on definite models cannot give a completely adequate description of all properties of real composites. Each case may present peculiar phenomena which either fall outside the scope of the models or make it necessary to modify them accordingly. For example, when analyzing the electroconductivity of composites, breakdown effects between filler particles may be encountered. In some cases the interface may be of great importance to conductivity and diffusion. The problem of active involvement of interfaces seems to have been least developed theoretically, so further intensive studies in this field may be expected. Certainly, specific phenomena associated with changes of the structure or properties of composites due to the effect of external fields, or the medium, should be borne in mind, namely breakdown (electric conductivity), changes of the properties, as well as stresses due to swelling, and effects arising from differences in the thermal response of the two phases.

2.5 Strength of Continuous Fiber-Reinforced Polymer Composites

In this Section we shall examine some approaches to the calculation of the strength properties of polymer composites. Although the practical significance of such calculation is obvious, the problem is so complicated that satisfactory calculation techniques are available only in a few individual cases.

When analyzing polymer composites, macroscopic and microscopic approaches may be distinguished. In the first case a composite is treated as a homogeneous anisotropic material. In the second case the material is analyzed as being heterogeneous, thereby making it possible to examine tensions in single fibers, filler particles, layers, etc. The microscopic approach permits to divide the process of failure into individual stages and to examine separately the processes of matrix-fiber delamination, fiber breakage, and destruction of matrix interlayers between filler particles.

The problems concerning the measurement, calculation, and prediction of the strength have been developed best for continuous fiber reinforced materials. This section is devoted to the main principles of the theory of the strength of such composites. The main attention will be centered on those aspects useful also for the analysis of the behavior of dispersion filled composites.

Composites reinforced with continuous twisted or straight fibers, braids, fabrics and bands are widely used to manufacture the structural components and load-bearing elements in the space technology and in the aircraft, automotive and shipbuilding industries. Therefore the measurements, calculations, and prediction of the strength of such materials are of great interest to engineers and researchers.

The main feature of a material of such type is a strong anisotropy of its properties, in particular of strength and elastic modulus. These composites are best described as a combination of anisotropic elements oriented in different directions and bound to each other by the polymer matrix. Several methods have been developed to calculate various properties, including the strength, of a product as a whole from the known characteristics of each of the anisotropic elements. A number of criteria of failure have been proposed [66–73] in which each element is treated as an anisotropic, but homogeneous material. Examples of such an approach are given in reviews [74–80]. Taken as a whole, this is called macroscopic mechanics of composites. However, the key problem of the material science for reinforced plastics is to calculate the properties of the anisotropic element, for example a unidirectional layer, on the basis of the known properties of the components, i.e. fiber and matrix.

The calculation of strength is far from being as accurate and rigorous as the solution of the linear elasticity problems. Besides, our present knowledge of the real structure of materials and actual mechanisms of failure is quite insufficient, which makes the calculation and prediction of composite strength, based on the properties of the components, even more uncertain. Nevertheless, it is in the field of fiber reinforced composites where the greatest progress has been made, which may be explained by a relatively simple mechanism of failure.

It is also important to emphasize the effects of stress-concentrating structural defects and inhomogeneities, such as fiber curvature, fiber breakage, pores, knots, etc., on the strength of a composites. We shall analyze the failure process and the ultimate strength, (i.e. the stress at which the bearing capacity of a specimen disappears completely) for unidirectional materials stretched and compressed along and across the reinforcement direction, and suffering a shear in the reinforcement plane. It should be noted that the properties of the composite components, i.e. fibers, matrix, and interface, have to be examined, making allowance for the actual structure in the stressed state.

2.5.1 Fiber-Aligned Tensile Strength of Unidirectional Composites

Let a specimen be stressed by a tensile load along the fiber orientation. In most of the polymer composites, the ultimate elongation of the fibers is considerable smaller than that of the polymer matrix. For example, $\varepsilon^* = 3–5\%$ for glass fibers, 0.8–3% for carbon fibers, 0.6–1% for boron fibers, and $\sim 4\%$ for organic (Kevlar type) fibers. The problem of the ultimate elongation of the polymer matrix (usually cross-linked three-dimensional polymers, i.e. polyethers, polyepoxides, hardened phenol-formaldehyde resins, etc.) is more complicated because ε^* for such brittle materials is essentially dependent on the preparation conditions and geometry of the specimen. The ultimate elongations for these binders, measured usually in massive specimens, is small (about 2–5%), However, in relatively thin film specimens ($< 100\ \mu m$) a considerable plasticity has been demonstrated; for example, $\varepsilon^* \simeq 10\%$ in the case of polyepoxide. In a composite material, where the matrix interlayer film thickness is even smaller by two orders of magnitude, the ultimate elongation of the matrix can be assumed to be at least 10%. Therefore, in the case of stretching of such specimens, fiber breakage will be the primary process of failure. In the further analysis of composite strength, it should also be taken into account that the reinforcing fibers are inhomogeneous and that there is a distribution of fiber strengths. In the case of brittle fibers (glass, carbon, boron) there are also defects on the surfaces and in the bulk of the fibers. The organic polymer fibers, on the other hand, have a complicated fibrillar structure (they are quasi composite microspecimens in themselves), and the mechanism of their failure is even more complicated. However, they also show a certain strength distribution. Thus, the fiber failure process is started by the breaking of the weakest fiber. The process may then proceed by at least two ways, namely, (a) fibers will break at their weak points irrespective of each other (i.e. the failure mechanism consists in the accumulation of independent breaks), or (b) breakage of one of the fibers will result in an overstress on its neighbor fibers, and they will be the next to suffer failure. In case (b), the failure process is localized, and initiates a turnpike crack which propagates in the material.

Obviously, higher material strength would be attained if the first mechanism of failure would be effective alone. In reality, however, the destruction process is initiated simultaneously by the two mechanisms. Independent breaks are accumulated initially and then a turnpike crack starts propagating. The question is how far the first stage of fiber failure may proceed. The longer the first mechanism prevails, the higher the ultimate strength that can be attained.

The value of the coefficient of stress concentration in the fibers adjacent to a broken fiber is the main criterion of the transition from one failure mechanism to the other. The higher this coefficient, the earlier a local failure begins with increasing load.

A number of theoretical and experimental works are devoted to analyzing this problem [81-86]. Any analysis should include static and dynamic effects. Special attention should be paid to the role of elastic energy released as a result of fiber breakage [87-88]. Although the relevant quantitative theory is far from being complete, some qualitative considerations and conclusions will be discussed here.

The following factors affect the static overstress coefficient.

(1) The larger the area of failure (e.g. the diameter of a broken fiber), the higher the overstress in such region.

(2) The coefficient of stress concentration depends on the matrix properties. If the absolute value of stresses in the region of fiber breakage is sufficiently high (as it is usually the case), the polymer matrix shows a plastic deformation in this region. The higher the ultimate plasticity of the matrix, the higher the stress concentration there. If elastic deformations prevail in the matrix, the stress concentration rises with increasing elastic modulus of the matrix.

Usually, only these two factors are analyzed in theoretical works, and the calculated stress concentrations turn out to be very high (values of 1.2–1.7). Indirect and direct experimental data often disagree with these estimates, indicating lower values. Probably, a crack initiated in the matrix by a fiber breakage may proceed further along the fiber, giving rise to a peeling of the matrix from the fiber near the broken region, thereby decreasing the stress concentration.

If the elastic modulus of matrix and fiber would be similar, the stress concentration at the point of fiber breakage would approach that of a homogeneous material.

In this case the main advantage of a composite would be lost and a turnpike crack would easily proceed across the fibers. This is why fibers and polymer organic matrices are generally selected such that the yield strength (usually below 100 MPa) and elastic modulus (below 5000 MPa) of the matrix differ from the corresponding fiber values by more than an order of magnitude.

The probability of local destruction increases [87–89] in the case of composites based on thick (~100–200 μm) glass, boron, or other fibers, especially if high modulus binders are used.

Hence, to estimate the upper limit of the strength of a unidirectional composite stretched along the fibers, the maximum stress resulting from the accumulation of independent breakages of fibers should be taken into account.

Let us examine now the strength of a fiber bundle. In order to find the strength of a bundle of parallel, unbonded fibers one has to know the fiber strength distribution function, or, more exactly, the ultimate elongation distribution. In this case, the elastic modulus of the individual fibers may be treated as a constant. Having analyzed experimental data, Weibull [90,91] proposed the following empirical expression for the fiber strength distribution, which has been widely used since to solve the problems relevant to the strength of composites:

$$F(\sigma) = 1 - \exp(-\alpha L \sigma^{\beta}) \tag{II.58}$$

where L is the gage, α, β are constants.
Figure II.6 shows schematically the relationship between fiber strength and the gage. Obviously, the larger the gage the higher the probability of a hazardous defect on this length. This fact determines the "scale factor" i.e. the decrease of the strength with increasing length of the fibers.

The strength of a parallel fiber bundle was calculated by Coleman [92] and Daniels [93] As a bundle is loaded, the weakest fibers break first. Because the weaker fibers are already broken and cannot support the bundle at maximum load, the bundle strength is lower than the mean fiber strength. The wider the fiber strength distribution, the

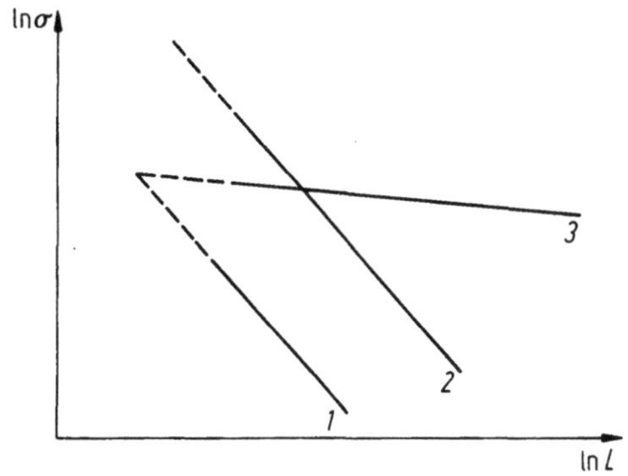

Fig. II.6. Schematic dependence of the strength on the gage; bundle (*1*), single fiber (*2*), and composite (*3*).

lower the bundle strength compared with the mean fiber strength. The bundle strength is given by

$$\sigma = (\alpha\beta Le)^{-1/\beta} \tag{II.59}$$

where e is the base of natural logarithms.
This expression does not contain the number of fibers in the bundle. It holds for bundles consisting of a great number of fibers. Commercial yarns generally consist of hundreds of fibers. For a small number of fibers, the strength depends upon this number, and more complicated expressions result. The dependence of the bundle strength on the gage L is shown in Fig. II.6.

Next a hypothesis permitting the prediction of the strength of a composite has to be developed. Once a fiber is broken, its end is no longer subjected to the tensile load. But, as in the case of composites with discrete fibers, a load is gradually transferred through the matrix to the broken fiber from its neightbors, due to shear stresses (see Section 1.1). Far enough away from the breakage point, the tensile load on the broken fiber appears to be the same as the average value for all fibers, i.e. the composite as a whole "forgets" the breakage.

Thus, there is only a small region in the specimen where the breakage is "remembered". This region is called the "ineffective length". Therefore, Gucer and Gurland [94] and Rosen [95] suggested that the strength of a composite should equal the strength of a fiber bundle of the ineffective length (it should be reminded here that the bundle strength depends on the gage). This suggestion is not based on any rigorous theory, but it has been substantiated to some extent by experimental data and mathematical simulation [96,97]. Then the composite strength expression is of the form:

$$\sigma_c = (\alpha\beta\delta e)^{-1/\beta} \tag{II.60}$$

where δ is the ineffective length.

The latter is the distance at which a load is transferred through the matrix to a broken fiber and may be calculated using Eqs. (II.19) and (II.25) described in Section 2.2.3. However, since we deal here with high stress levels, the equation for a plastic matrix (Kelly-Tyson [42]) may be used instead:

$$\delta = \frac{\sigma_f}{2\tau_m} \cdot d \tag{II.61}$$

Before analyzing this expression it should be noted that Daniels [93] assumed that all defects in all fibers are distributed randomly. (The validity of this assumption will be discussed below.) Such a model predicts independence of the composite strength on the gage.

This latter condition is almost satisfied in experiments where filaments are first impregnated with a polymer binder and then hardened (example 3 in Fig. II.6). The expression II.61 is also valid when stress is transferred due to friction between the matrix and the fibers. In this case τ_m should be replaced by the frictional stress (friction force divided by area), which may or may not equal the matrix-to-fiber adhesion strength. The adhesion strength measured, for example, as the force with which a fiber is torn out from the matrix, does not always equal the friction force; therefore expression (II.61), with "adhesive strength" replacing τ_m, must be used with caution.

For an elastic matrix, i.e. in the case where the stress is mainly transferred within an elastic region of matrix activity, δ may be obtained from the following equation [45]:

$$\frac{\delta}{d} \approx C\left[\frac{1 - v_f^{1/2}}{v_f^{1/2}} \cdot \frac{E_f}{G_m}\right]^{1/2} \tag{II.62}$$

where C is a constant close to 1.0.

Equations (II.61) and (II.62) show that, according to the described model, a matrix with the highest possible value of shear modulus G and shear cohesion strength τ should be chosen. Then the ineffective length would be minimum, and the strength of the composite maximum.

Figure II.7 shows how the composite strength depends on the parameter β of the Weibull distribution. Note that an increase of β^{-1} means a narrowing of the distribution (cf. Eq. (II.58)).

For two values of β, the dependence of the relative strength of a composite on τ_m and G_m is shown in Fig. II.8, as calculated using Eq. II.60, with δ obtained either from Eq. (II.61) or (II.62). It should, however, be born in mind that an increase of shear modulus and shear adhesion strength of the matrix is usually accompanied by an increase of brittleness and a decrease of the ultimate elongation of the composite.

From the viewpoint of the described model one should increase the adhesion in order to preclude a peeling of the matrix from the fiber at the point of fiber breakage, and increase the ineffective length. However, an excessive rise of the matrix adhesive strength and rigidity may lead to an adverse result in that the stress concentration coefficient would increase, and the "local" destruction mechanism would prevail.

Organoplastics, i.e. composites consisting of polymer matrices and organic fillers, are of special interest because the organic high-strength and high-modulus fibers are delaminated under breakage, due to their high anisotropy and fibrillar structure. Therefore, in this case the ineffective length is defined rather by fiber properties and by the length and form of the delaminated region of the fiber, than by matrix properties.

As a result, the strength of organoplastic specimens is practically independent of the properties of the matrix.

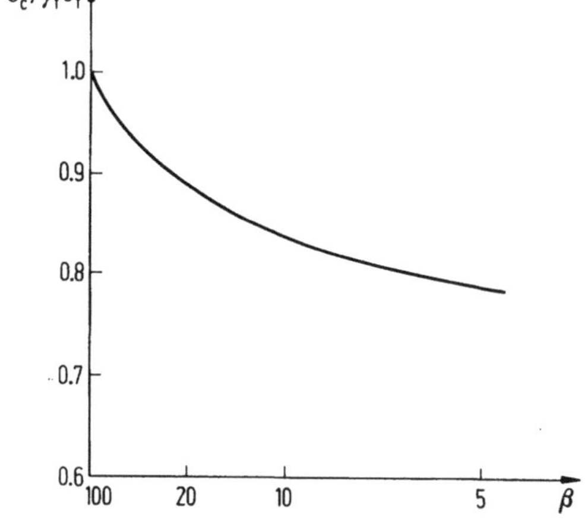

Fig. II.7. The dependence of the unidirectional composite strength on the parameter β of the Weibull distribution (calculated from Eq. II-60, all other parameters constant).

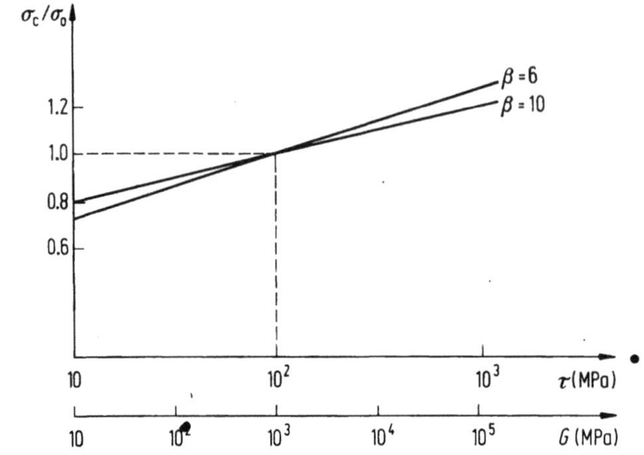

Fig. II.8. The dependence of the relative strength of a unidirectional composite on the shear modulus G and the shear cohesion strength τ of the matrix, for two values of the parameter β of the Weibull distribution; $\sigma_0 = \sigma_c$ at $\tau = 10^2$ MPa ($G = 10^3$ MPA), $d = 70$ μm, $\sigma_f = 4$ GPa, $c = 1.0$. $\gamma_f = 0.7$ and $E_f = 70$ GPa; α = const.

2.5.2 Factors Restricting the Application of Simple Models for the Failure of Composites

The Rosen-Daniels model has been based on the Weibull distribution of single-fiber strength. The applicability of such model is limited by two circumstances. In some cases the fiber strength distribution is different. For example, the strength distribution of single organic kevlar fibers is described by a bimodal curve [98]. In this case the strength of the composites has to be calculated making allowance for the real distribution function. Besides, the measurements of single-fiber strength are generally made on test samples (10 mm and longer) which are longer than the ineffective length, (fractions of one mm and smaller). Also other defects may be present in the fibers of a test sample, compared with the same fibers in a composite. Glass fibers [99] may be mentioned as an example. In this case the Weibull distribution parameters inferred from measurements of single-fiber strength cannot be used to calculate the strength of a composite. Besides, it should be remembered that all the above reasonings are valid for the case of brittle fibers (i.e. fibers showing a linear stress/strain diagram). This is the fact for glass, carbon, and boron fibers, whereas organic polymer fibers, and some of the metallic fibers, show a nonlinear elasticity or plasticity. Therefore, the model described above should be applied cautiously to the latter fibers.

The model also assumes a random distribution of single fiber strength in a yarn. In reality, however, the very process of yarn formation, i.e. the simultaneous extraction of many filaments from a spinneret, and the effects of various technological factors, will give rise to a definite variation of the strength of single fibers within a yarn cross section. As a result, the filament strength shows a significant variance which proves to be much in excess of the relevant theoretical value inferred from the Daniels model (the theoretical value of variance is inverse to the square root of the number of fibers in a yarn [93]).

Moreover, the yarn strength varies along the yarn length. There is also a certain dependence of the composite strength on the gage (line 3 in Fig. II.6), and a scale factor arises for products made of such composites (in large products, of course, there are other reasons for a scale factor, among them larger defects in their structure).

Technological imperfections in the preparation of the reinforcing materials result also in the fact that the Weibull distribution parameters α and β can be different in different batches of yarn, in different yarns of a tow, in different batches of tows, etc. Such inhomogeneities and variations of the reinforcing materials must be allowed for when estimating the strength of a composite. In this case the failure mechanism may change from successive random ruptures of individual fibers to a correlated simultaneous failure of a group of fibers (yarn as a whole). This means that such a composite should be treated as a combination of successively ruptured yarns rather than of single fibers, with the following consequences: First, this circumstance gives rise to a decrease of the strength of a composite even if the yarn ruptures are independent, i.e. a model of yarn rupture accumulation may be applied to a given composite. Second, an increase of the diameter of an element under failure (a yarn instead of a single fiber) results in an increase of the ineffective lengths (see Eqs. II.61, 62). Third, the increase of the diameter leads to an increase of the stress concentration coefficient in the neighboring fibers, i.e. to a higher probability of local failure, and hence to strength deterioration of the composite.

Let us now discuss in some more detail the concept of the ineffective length. Several factors may result in a (true or apparent) increase of this parameter. First, the shear analysis used to derive Eqs. II.61 and II.62 is not sufficiently rigorous and gives rise to errors. The problem of stress transfer from matrix to fiber has been solved [100], using a more rigorous method of mechanics, and showing that the tensile stress in a fiber may vary nonmonotonically along the distance from the fiber end, going through several maxima. This gives rise to secondary fiber-aligned ruptures, and has the same effect as an increase of the ineffective length. Multiple rupture in a single fiber was in fact observed experimentally [101]. Finally, multiple rupture of a fiber may also ensue from a dynamic effect, namely from a rapid release of elastic energy. Since fiber rupture is rapid and the matrix suffers high loading rates at the moment of fiber rupture, both matrix and adhesion are very brittle; a rapid development of a brittle crack is posible, either in the matrix along the fibers, or at the fiber-matrix boundary, thereby also resulting in an increase of the ineffective length. In fact, when studying the failure of model specimens of glass-reinforced plastics with a polarizing microscope, we observed two characteristic ineffective lengths, one of which was obviously due to delamination of the matrix from the fiber [101]. Leksovsky [87] believes that the release of elastic energy as a result of fiber rupture (especially if this energy is high) may increase the probability of the rupture of neighboring fibers and give rise to a correlated simultaneous rupture of a group of fibers. Although quantitative substantiation of this hypothesis [87] is not quite convincing, one should bear in mind that such a process might be possible. On the contrary squeezing which results in an increase of the matrix yield strength, reduces the actual ineffective length compared with its value calculated, for example, by the Kelly-Tyson formula [102].

2.5.3 The Role of Structural Inhomogeneities and Defects

Consider now the structure inhomogeneities and defects of a material which may affect its strength. When manufacturing a product from a composite material, the difference in stretching and/or length of individual fibers, yarns, tows, etc. are cha-

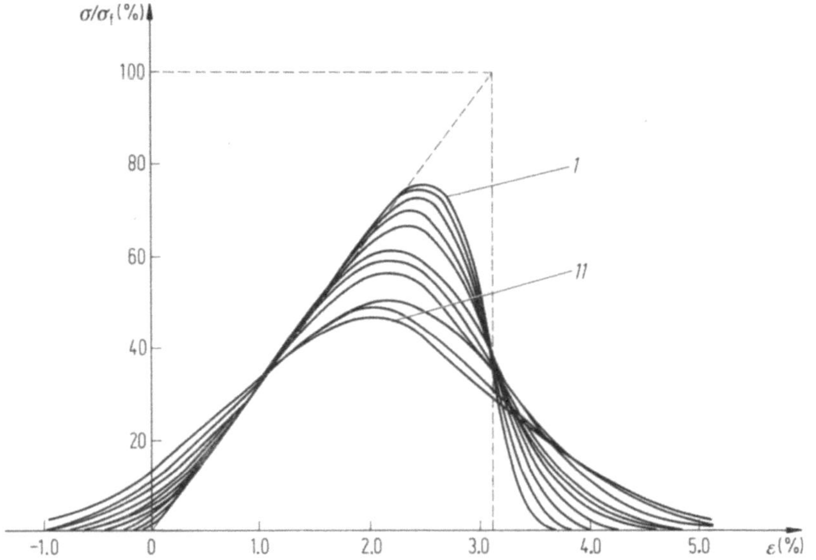

Fig. II.9. The effect of the dispersion of length on the glass fiber bundle stress-strain diagram, where $\bar{\sigma}_f$ is the average strength of the single fibers. From curve *1* to curve *11* the dispersion of distribution of the fiber length increases from 0 to 2% · 104).

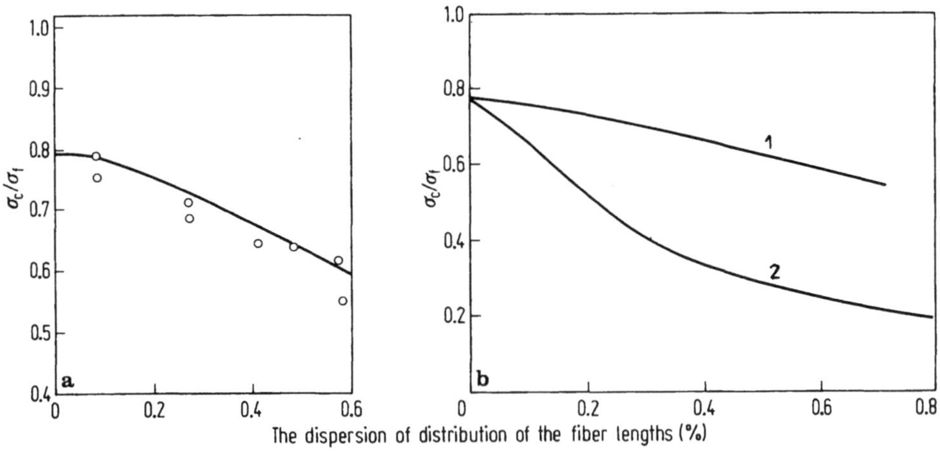

Fig. II.10. a, b. The effect of the dispersion of distribution of the fiber length on the composite strength (104). **a** continuous curve: calculation; points: experimental data for composites based on Kevlar fibers; **b** calculated curves for composites based on glass fibers (*1*) and carbon fibers (*2*).

racteristic inhomogeneities. Such difference results in a non-synchronous action of various reinforcing elements and, consequently, in an expansion of the load (deformation) interval where failure occurs; this leads eventually to a deterioration of the product strength. The effect of the distribution of fiber and yarn lengths in a composite has been studied theoretically and experimentally [103,104]. The calculated effect of the dispersion of distribution of the fiber lengths on the fiber bundle stress-strain diagram is shown in Fig. II.9 for glass fibers. The effect of the same parameter on the maximum stress (strength) of a unidirectional specimen of a composite is shown in Fig. II.10a. The effect of the dispersion of distribution of the fiber lengths should

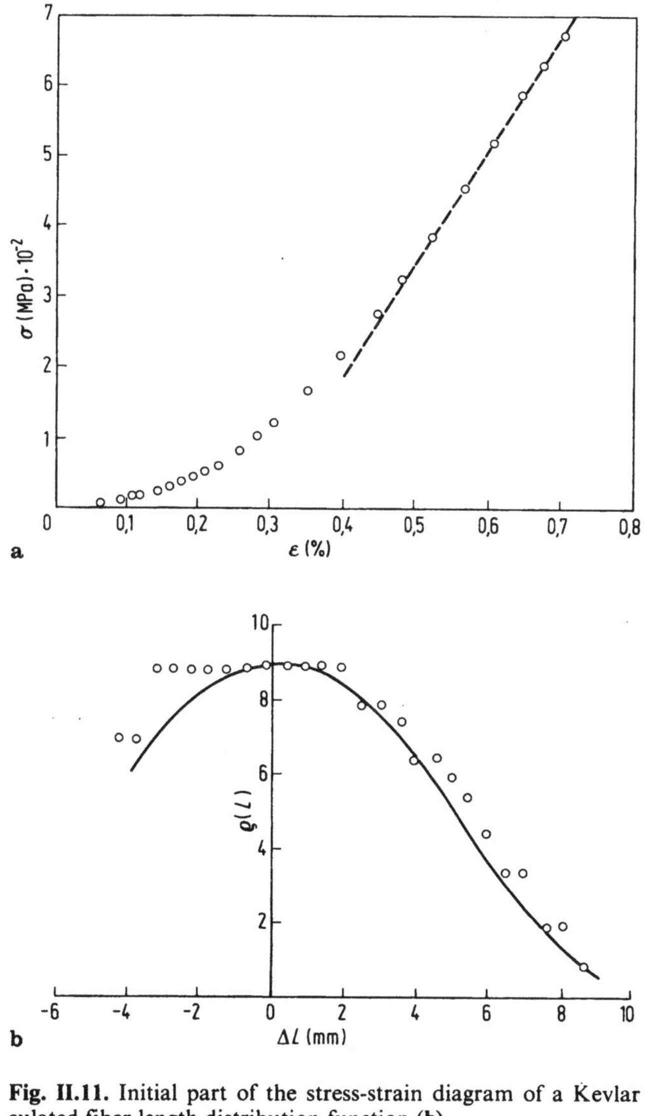

Fig. II.11. Initial part of the stress-strain diagram of a Kevlar fiber bundle (**a**) and the calculated fiber length distribution function (**b**).

increase with an increase of the elastic modulus of the fibers (with decreasing ultimate strain). Figure II.10b presents the calculated curves of the strength decrease with increasing dispersion of distribution, for composites based on glass and carbon fibers. In the latter case a much stronger effect can be seen. Unfortunately, no methods for the experimental quantitative analysis of the variability of lengths and tensions of reinforcing elements in real products have been developed as yet, making it difficult to use these theoretical approaches. We want to mention, however, that variable lengths and tensions of fibers do appear when reinforcing materials (yarns, tows, bands, fabrics, etc.) are manufactured, and when such materials are reprocessed into products, during their reeling, piling, etc. Obviously, technological methods which may reduce the length and tension variability should be used. Of course, this problem, like many others, must be solved with a view to economic optimization, bearing in mind that a reduction of the length variation increases the strength.

An experimental method of estimating the fiber length variability in reinforcing elements consists in analyzing the initial segments of the stress-strain diagrams [104]. The method is based on the assumption that the slope of the σ-ε diagram at each point reflects the number of acting fibers. Figure II.11 shows the stress-strain diagram of Kevlar fiber yarn, and the calculated fiber length distribution function (curve b is obtained by graphic differentiation of curve a).

The deviations of fiber orientation or, which is the same, the variations of the stress direction during tests or utilization of a product, constitute another type of structural inhomogeneities of a composite material. In this case, the material suffers transverse and shearing tensile stresses which may give rise to delaminations in matrix, interface, or fiber (this is typical for anisotropic organic polymer fibers).

A unidirectional composite is known to be extremely sensitive to the direction of load application. This effect is most striking in organic polymer fibers due to the low shear strength of these fibers. As an example, the polar strength diagrams for unidirectional specimens of glass, and organic polymer fibers are shown in Fig. II.12 [105].

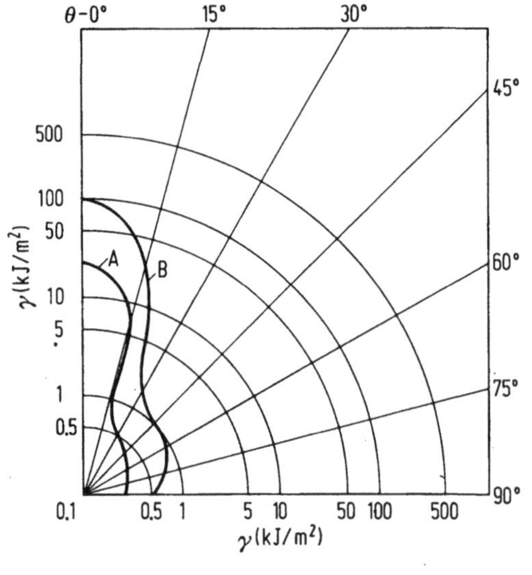

Fig. II.12. Polar diagrams of the effective surface energy of the composite fracture, γ. A: Kevlar fibers; B: glass fibers (105).

The delaminations resulting from fiber disorientation lead, in the case of uniaxial tension, to an apparent increase of the ineffective length, to a decrease of strength, and also to a significant loss of bearing strength under other stressed states (bending, compression, etc.).

The problem of a quantitative inclusion of the disorientation effect is difficult and has not been solved definitely. The following equations may be used to estimate the strength of composites with curved or disoriented fibers:

$$\sigma_c^{\varphi} \approx \frac{\sigma_c^0}{1 + 2\dfrac{\sigma_c^0}{\sigma_c^{\perp}}\varphi^2} \qquad \varphi < 10^\circ \text{ for monophase curvings } ^{106)} \qquad \text{(II.63)}$$

$$\sigma_c^{\varphi} \approx \frac{\sigma_c^0}{1 + \left[\dfrac{E_f}{G_m} v_f(1 - v_f) - 2\right]\varphi^2} \qquad \varphi < 10^\circ \text{ for antiphase curvings } ^{107)} \qquad \text{(II.64)}$$

Experimental data [105] agree with the estimates obtained using these equations.
Regular curvings are characteristics of materials based on twisted filaments and on fabrics, while irregular curvings result from imperfections of reprocessing into a composite material.

In a material with fiber disorientation, the strength deterioration increases with enhanced anisotropy of the composite. The strength is much affected by extended defects, i.e. structural failures across the fibers. Such defects are caused by ruptures of weak filaments, tows, and knots, by inclusions, pores, etc. The deterioration of strength of a composite with such defects is due to stress concentration on nonruptured fibers near the defects. A similar problem was solved for homogeneous materials as early as in the twenties [108, 109]. As a result the famous Griffith formula for the effect of crack size and properties of a material on its strength was obtained:

$$\sigma \approx \left(\frac{2\gamma E}{\pi c}\right)^{1/2} \qquad \text{(II.65)}$$

where c is the crack size.
The corresponding problem for a composite material with a plastic matrix was solved theoretically [110, 111] showing that the dependence of the composite strength on defect size is weaker than in the Griffith formula and can be described by the logarithmic law*

$$\sigma_c = \sigma_c^0 - q \ln\left(\frac{C}{C_0}\right) \qquad \text{(II.66)}$$

* Here and in the following we use the expression obtained in ref. [111]. The law of strength variation with the size of the defect obtained in ref. [110] is the same, but the coefficients are different. However, we were not able to estimate its applicability and validity because the source where the derivation of the formula was published was not available to us.

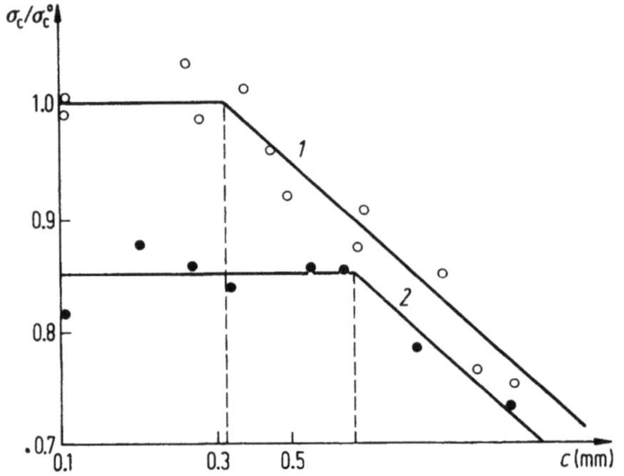

Fig. II.13. The dependence of the unidirectional composite strength on the size of defect [111]. Curve *1*: Kevlar, tow d, curve *2*: Kevlar tow d_2, $d_2 \simeq 4d_1$.

where σ_c is the strength of a composite with a notched linear defect of length C (the problem has been solved only in the one-dimensional approximation); σ_c^0 is the strength of the unnotched specimen; C_0 is the internal defect size determining the strength of the unnotched composite; q is a factor characterizing the sensitivity of the composite to the defect.

Equation (II.66) was experimentally verified, using a Kevlar and a glass-fiber reinforced composite. The defects were tows of various dimensions made of a material with a smaller ultimate elongation than that of the basic filler fibers [111]. The applicability of Eq. (II.66) at $C > C_0$ was confirmed. At $C < C_0$ the strength remained constant $\sigma_c \simeq \sigma_c^0$ (see Fig. II.13). Moreover, C_0 has proved to be determined by the size of the reinforcing element (or by the tow diameter).

The value q characterizing the sensitivity of a material to defects is given by [110,111]:

$$q = \frac{2\tau_m}{\pi}\sqrt{\frac{E_c}{G_c}} \approx \frac{2\tau_m}{\pi}\sqrt{\frac{E_f v_f(1 - v_f)}{G_m}} \qquad \text{(II.67)}$$

Thus, high-modulus composites based on, for example, carbon and boron fibers should be more sensitive to defects, and this is in fact usually observed experimentally. To reduce the sensitivity of a composite to defects, one should also choose a matrix with a low shear strength τ_m. This latter appears counterproductive, since high strength is generally aimed at. However, the larger the defects in a given filler, the greater may be the relative advantage of using a matrix with low yield strength. In order to obtain quantitative estimates of q and correspondingly, of σ_c from Eqs. (II.67) and (II.66), it should be taken into account that τ_m in a composite may differ from τ_m measured in free polymer specimens due to squeezing effects, the physicochemical influence of the fibers, etc. Besides, a real polymer matrix is usually not an ideal elastoplastic body.

The above analysis dealt with defects of a comparatively small size ($C < 5$ mm). In the case of stress concentrators of much larger size, the conventional failure criteria of linear mechanics are quite satisfactory, i.e. the strength is inversely proportional to the square root of the defect size [112,113]. This is associated with the fact that the size of the plastic deformation zone at the point of the highest concentration of stress (crack mouth) is independent of the size of the defect. The transition from the rule of logarithmic type (formula II.66) to the Griffith-type (Eq. II.65) was in fact observed experimentally [111].

Residual stress arises as a result of the matrix hardening process or, more exactly, of the final stages thereof, during the cooling from the hardening temperature, when the matrix material is already in the glassy state. Numerous theoretical [114–116,133] and experimental [117–119,122] works have been devoted to analyzing residual stresses and their effect on the strength of a product. Here, as in other problems relevant to composites, macroscopic and microscopic approaches may be distinguished.

Apart from filler and matrix, a composite material contains pores, resulting from the impregnation and, mainly, from the hardening of the matrix. The pores have a significant adverse effect on the compressive and shear strength (see below), but affect little the tensile strength of unidirectional specimens, along the fibers. If fibers located inside a pore are considered to behave as a bundle of unbounded fibers, the strength of such a region in a composite will depend on the geometric dimensions of the pore. The strength of the material in the pore will be much below the strength of the basic material of the composite only, if the dimension of the pore along the fibers is much in excess of the ineffective length, and the cross section permits a great number of fibers to be housed in. After the failure of the fibres in a pore, the latter becomes a defect in the material, and the deterioration of the strength should be analyzed further by an equation like Eq. (II.65). However, the existence of such large pores is unlikely in real composites and, therefore, the effect of porosity on the tensile strength along the fibers seems to be insignificant.

2.5.4 Transversal Strength of Composites

The products made of composite materials and working under the simplest state of stress are shells and pipes resisting an internal pressure. They are usually made of unidirectional composite layers in such a way that each of the layers suffers a biaxial tension. Therefore, the strength of a composite under transverse tension (so called

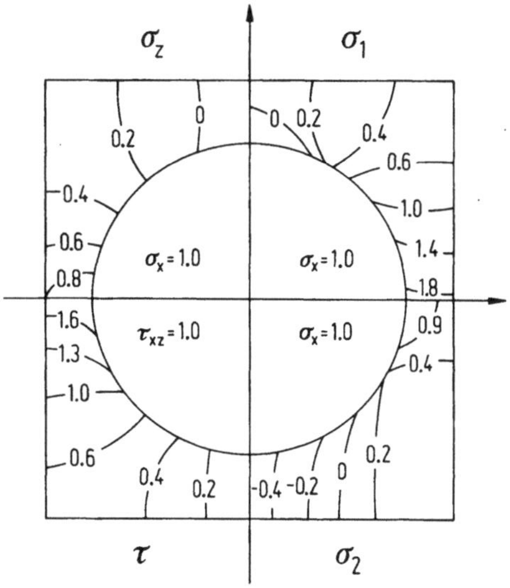

Fig. II.14. Stress fields in unidirectional composite based on boron fibers (122).

transversal strength) is of interest. Since the ultimate transversal strength is most frequently determined by plastic shear deformation in the matrix, it is useful to simultaneously examine the compressive strength across the fibers and the shear strength in the fiber plane. The stress fields in a unidirectional specimen are complicated and inhomogeneous. They are usually analyzed in terms of regular models of composites using numerical calculational techniques, for example the finite-element method [38,39,120,121]. Figure II.14 shows an example of a numerical calculation of the stress fields in a fibrous unidirectional composite based on boron fibers; σ_1 is the stress along the load direction, σ_2 that across the load direction, and σ_Z that along the fiber direction; τ is the shear stress in the XZ plane. The values on the curves are the level of the relative stress in these curves. There is an overload zone in the matrix where $\sigma_1 > 1.0$ and $\tau > 1.0$ [122]. Under stress, the failure process will start just in this zone. The stress concentration coefficients will be the main parameters characterizing the strength of such a material. The values of the concentration coefficients of tensile and compressive stresses, $K^{\pm}$, and of shear stress, K^{τ}, which characterize the maximum-to-mean stress ratio, have been calculated [123]. The calculations were made disregarding any difference in Poisson's ratios and, what is more important, the problem has been solved only for a purely elastic deformation. The approximate expressions for the stress concentration coefficients in a composite with square packing of fibers are of the form

$$K^{\pm} = \frac{1 - v_f\left(1 - \dfrac{E_m}{E_f}\right)}{1 - \left(\dfrac{4v_f}{\pi}\right)^{0,5}\left(1 - \dfrac{E_m}{E_f}\right)} \tag{II.68}$$

$$K^{\tau} = \frac{1 - v_f\left(1 - \dfrac{G_m}{G_f}\right)}{1 - \left(\dfrac{4v_f}{\pi}\right)^{0,5}\left(1 - \dfrac{G_m}{G_f}\right)} \tag{II.69}$$

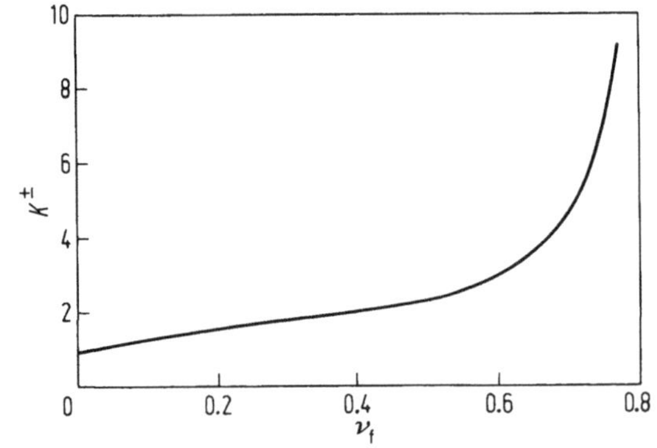

Fig. II.15. The dependence of the stress concentration coefficient $K^{\pm}$ on the degree of filling, according to Eq. II.68 (for $E_f \gg E_m$)

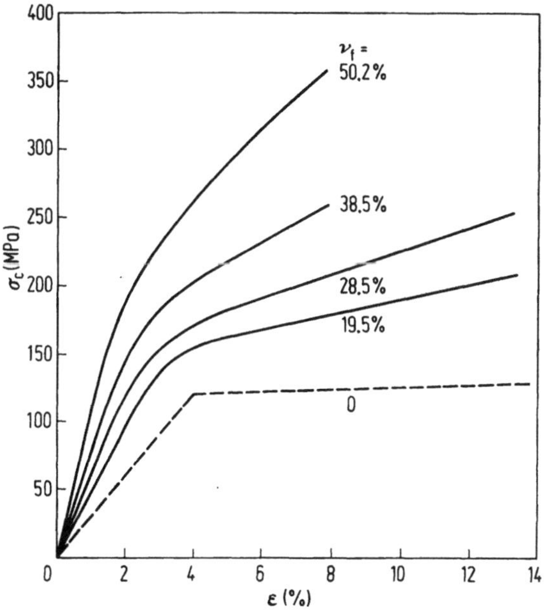

Fig. II.16. Calculated stress-strain diagram for a transverse tensile load on a unidirectional composite [101].

Fig. II.15 shows $K^{\pm}$ as a function of filling at $E_f \gg E_m$. For the assumed failure mechanism, the tensile, compressive, and shear strengths are given by:

$$\sigma_c^+ = \frac{\sigma_m^+}{K^{\pm}} \quad \text{(II.70)} ; \qquad \sigma_c^- = \frac{\sigma_m^-}{K^{\pm}} \quad \text{(II.70')} ; \qquad \tau_c = \frac{\tau_m}{K^{\tau}} \qquad \text{(II.70'')}$$

where σ_m^+, σ_m^-, τ_m are the tensile, compressive, and shear strength of the matrix. Since v_f is generally 0.5–0.7, $K^+ \approx K^{\tau} \approx 2 \div 7$, and the transversal strength of a composite under interlaminar shear is 1/7 to 1/2 of the respective strength of its matrix. Generally speaking, this value is not too far from experimental data despite very rough assumptions in the derivation of the equations for stress concentration, and in the adopted failure mechanism. A composite may in fact be assumed to suffer a failure when a critical deformation rather than a critical stress in its matrix is attained; this, however, will little affect the quantitative estimate because the equations for the concentration coefficients of deformation are similar to those of stress [123].

The results of the more accurate calculations of the stress and deformation fields by the finite-element method [121–124] are not much different from these approximate estimates. It is our opinion that the solution of the elastoplastic problem (as, for example, in ref. [129]) will be the most important step in further refinement.

A. B. Givental [101] calculated the diagram of transversal tension (compression) of a composite making allowance for an elastoplastic behavior of the matrix. The ultimate plasticity of the composite is close to that of the matrix, or may even exceed the latter (see Fig. II.16). In other words, the elastoplastic problem leads to a smoothening of the stress fields, and

$$\sigma_c^+ \approx \sigma_m^+ \approx 2\tau_m \qquad \text{(II.71)}$$

Thus, the transversal and interlaminar strength of a composite appear to be somewhat underestimated in the pure elastic approximation.

Considering a reduction of stress concentration, is may appear interesting to prepare a material with thin uniform matrix layers (as mentioned in Chapter I). However, this idea is hardly realizable in practice. Such a material would require a broad distribution of fiber diameters, and the fibers would have to be distributed spatially in quite a definite way. Another possibility would be a variation of the shape of the fiber cross section [126]. This may make it possible to increase the filling, and may also be useful for improving the transversal properties of composites. However, thus far this method has been realized only for polymeric (i.e. highly anisotropic) fibers with low transversal strength.

In order to estimate the serviceability of composite products in vessels under biaxial tension, the transversal ultimate strain, rather than the transversal strength, is of importance, because the former is determined by the axial ultimate strain of the other layers. If the transverse ultimate strain is smaller than the axial strain, the final failure is preceded by delamination with fiber-aligned cracks in all layers. In this case the vessel gets depressurized and very compliant to bending, loses its stability to load variations, etc. In glass- and carbon-fiber reinforced plastics, the transversal ultimate strain may be increased by improving the matrix-filler adhesion strength, and by choosing a matrix with a high ultimate strain. The required ultimate strain of the matrix may be estimated by the formula

$$\varepsilon_m^* > \frac{\bar{\varepsilon}_f^* K_d}{v_m} \tag{II.72}$$

where K_d is a strain concentration coefficient; $\bar{\varepsilon}_f^*$ is the fiber bundle ultimate strain along the ineffective length. The difficulty is to find the value of K_d, because this involves calculation of the strain fields in the region where the matrix behaves explicitly nonlinearly. Besides, as indicated above, the behavior of a matrix in its free state and in a composite may be different due to squeezing and other effects. Difficulties which have not been overcome as yet arise also when estimating the necessary value of the adhesion strength. Here again, plastic matrices filled with organic fibers constitute a special case. The failure under transversal tension and shear seems to be initiated [127–129] by the failure of the fibers which are the weakest point. Therefore, the delamination and depressurization of loaded "organoplastic" vessels take place already at pressures which are only 10–20% of the failure pressure. Since in this case the fibers play the leading role, the situation cannot be altered much be selecting a more plastic matrix.

The transversal properties of an unidirectional composite are probably little sensitive to structural inhomogeneities such as small curvings and disorientation of the fibers. However, studies on that subject are not known to us. On the other hand, the effect of pores, inclusions, and other stress concentrators must probably be examined in direct relation to the ability of a composite to withstand the propagation of cracks along the fibers, in the matrix and at the interface. The shape of pores and inclusions, their fiber-aligned dimensions and effective curvature radius, i.e. the factors defining their degree of hazardousness, are important in this context. Approaches based on linear failure mechanics [130,131], appear to be most promising.

It is obvious that the effect of structural defects on the transversal strength and interlayer shear will be reduced if the adhesion strength against breakoff and shear are increased and the failure viscosity of the matrix and its plasticity are improved. These circumstances have led to an extensive usage of epoxy matrices modified with rubber. The fiber distribution in the composite cross section (agglomeration, etc.) may, and probably does, affect the transversal ultimate properties. First, such agglomeration may completely remove the binder (matrix) from certain zones of fiber contact. In this case the danger of failure increases, i.e. the strength of the brittle fibers deteriorates, pores appear, etc. Second, a nonuniform distribution gives rise to regions of increased brittleness, whence the failure process originates. These problems have been little studied. They are also characteristic of particle-filled composites, and we discussed them in Section 1.3.1.

2.5.5 Compressive Strength

Another utilization of composites is in vessels subjected to external pressure and in other elements suffering compression. In this case the strength of a unidirectional specimen of a composite under axial tension is an important parameter. In homogeneous plastic materials, the mechanism of losing bearing ability under tension and compression is usually the same, namely a shear at about 45° to the loading direction. Therefore the tensile and compressive strengths of such materials (also metals) are similar. Homogeneous brittle materials suffer failure under tension usually through the development of a crack into a break, whereas a shear crack propagates in case of compression. Correspondingly, such materials are characterized by much higher values of compressive strength compared with tensile strength.

The mechanisms of failure of unidirectional composite materials under tension and compression along the fibers are quite different. The axial compression strength may be either much lower (organoplastics) or higher (boroplastics), than the tension strength.

The failure of unidirectional composites under axial compression is commonly believed to be due to stability loss. This problem has been treated theoretically [132-137]; unfortunately it cannot be considered as well defined, not even for perfect unidirectional structures. Two forms of stability loss, viz. symmetric where all fibers bend in the same phase, and antisymmetric where all fibers bend in the opposite phase, have been examined [132,133,137]. For the symmetric one, the following expression for the compressive strength (stability loss stress) was obtained, in the particular case of elastic behavior of fibers and matrix [133]:

$$\sigma_c^- = 2v_f \left[\frac{v_f E_f E_m}{3(1 - v_f)} \right]^{1/2} \tag{II.73}$$

If, however, the matrix behaves as an elastoplastic body and plastic deformation plays the leading role in failure, the expression for the compressive strength takes in following form [132]:

$$\sigma_c^- = \left[\frac{v_f E_f \tau_m}{3(1 - v_f)} \right]^{1/2} \tag{II.74}$$

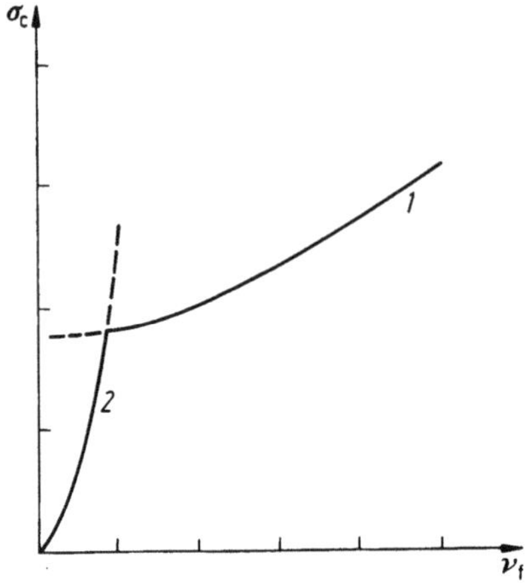

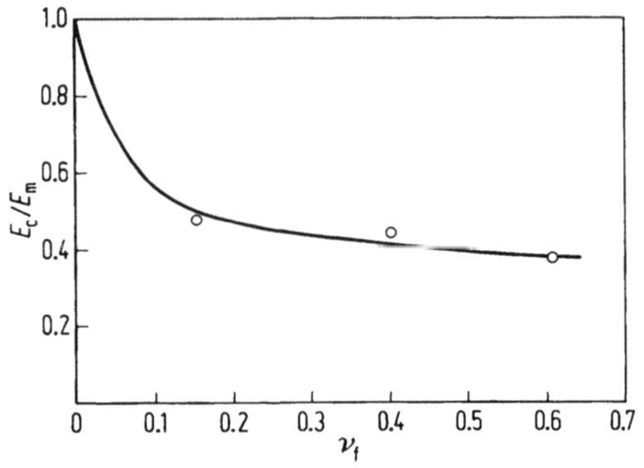

Fig. II.17. The dependence of the compressive strength on the volume fraction of fiber according to Eqs. II-73 and II-75. *1*: symmetric form of stability loss; *2*: antisymmetric form of stability loss.

Fig. II.18. Dependence of Young's modulus of HDPE composites filled with glass spheres, on the volume fraction of the filler modified by liquid oxyalkylene organosiloxane oligomer; $d_f = 30$ μm, $C_{olig} = 1\%$ of filler.

For the antisymmetric form of stability loss, and elastic behavior of the components, the following equation was derived [133]:

$$\sigma_c^- = \frac{G_m}{1 - v_f} \qquad \text{(II.75)}$$

Figure II.17 (obtained from Eqs. (II.73) and (II.75)) shows qualitatively that at low volume fractions of fiber the strength is limited by the antisymmetric, at high v_f by the symmetric form of stability loss.

Equations (II.73), (II.74), and (II.75) indicate that the axial compressive strength is independent of the fiber diameter. In fact, mechanics teaches that the load P, at which an elastic rod in an elastic medium loses its stability, is proportional to d^2 (d = diameter of the rod) [138], whereas the stress, which is equal to $P/\pi d^2$, is independent of the rod diameter [138]. However, some experimental data of composites based on glass and boron fibers indicate a certain, though small, rise of strength with increasing fiber diameter, at axial compression [105,139-142]. Such a rise may perhaps be due to a better (closer to ideal) structure of composites based on thicker fibers in which curvings, disorientations, etc. are less pronounced. However, the compressive strength of glass-fiber reinforced plastics decreases somewhat in the range of high fiber diameters (150–200 μm)[139-141]. At the same time, the axial tensile strength of such composites decreases monotonously with increasing fiber diameter. This is probably due to a scale factor, i.e. to an enlargement of the fiber surface and, consequently, to a rise of the probability of hazardous defects per unit fiber length. The decrease in the compressive strength of glass-fiber reinforced plastics with large-diameter fibers seems to correspond to a change of the failure mechanism, from stability loss to fiber failure. In the latter case the strength of the composite should be

$$\sigma_c^- = v_f \sigma_f^- + v_m E_m \varepsilon_f^{*-} \tag{II.76}$$

where σ_f^-, ε_f^{*-} are the strength and ultimate compressive strain of the fibers (on the assumption that the ultimate compressive strain of the matrix is higher than that of the fibers). A peculiar mechanism of failure under axial compression is characteristic of composites based on Kevlar fibers [127,129,143]. When compressive loads are applied, the organic polymer fibers lose their stability, at a certain critical load, in a peculiar manner. They form folds of the type typical of polymeric single crystals. This phenomenon is caused by the high anisotropy of the organic fibers and by their weak limiting transversal and shear parameters. In this case the compression strength of a composite is described by expression (II.76).

Equations (II.73), (II.74), and (II.75) have been derived with the assumption that the matrix is either elastic or plastic, up to the moment of failure of the composite material. It was also assumed that the adhesive bond at the fiber-matrix interface remains intact until the moment of failure. In the particular case of carbon fiber reinforced-plastics, it was shown [144–146] that an increase of the adhesive strength (by special surface processing or whiskerization of the fibers) simultaneously improves the interlaminar shear strength, the transversal tensile strength, and the axial compression strength. In this case, obviously, the break of adhesion is the limiting stage of failure, and the compressive strength may be calculated using Eq. (II.74), but substituting shear strength of adhesion for shear strength of the matrix, τ_m.

Structural inhomogeneities and defects will significantly affect the axial compression strength. Delamination along fibers under compression, which reduces the load necessary for a material to lose its stability, seems to be the most general consequence of inhomogeneities and defects. The effect of fiber disorientation and curving on the strength of unidirectional composites has been studied experimentally [123,147]. The effect of pores on the compressive strength was examined theoretically [121,124], on the basis of the mechanism of antisymmetric stability loss, and the following equation was proposed:

$$\sigma_c^- = \left(\frac{G_m}{1 - v_f + v_f \dfrac{G_m}{G_f}} \right) \left(\frac{1 - \dfrac{2v_v}{1 - v_f} + \dfrac{v_v^2}{(1 - v_v)^2}}{1 + \dfrac{v_v}{1 - v_v}} \right) \tag{II.77}$$

where v_v is the volume fraction of pores in the matrix. The theoretical model disregards size and shape of the pores, because it fails to treat a pore as a stress concentrator. However, some available data indicate an effect of shape and size of the pores. It seems to us that in the future attention should be paid to the theoretical and experimental analyses of the delamination mechanism in composites under axial compression, and of the effect of various defects as stress concentrators, on the value of the strength.

2.6 Experimental Results and the Reasons for the Disagreement Between Theoretical Results and Experimental Data

This chapter so far dealt with the theoretical models and computation methods for the values of mechanical properties of various types of polymer composites. Now we are going to compare theoretical and published experimental results.

Experimental values of Young's modulus have been published for many types of composites. For example, for dispersely filled composites filled with glass spheres, at filler volume fraction 0.10–0.50, E_c values were reported [148–156]. Various polymer matrices have been used in these papers: three-dimensional crosslinked networks (polyurethane [154]), semicrystalline thermoplasts (HDPE [148,149], polypropylene [152], polyamide [155]), amorphous thermoplasts (poly(methyl metacrylate) [150], polystyrene [153], poly(vinyl chloride) [156]). Most of the experimental results agree quite well with the computations based on the models described in Section 2.1 for $v_f \leqq 0.3$. The relative increase of Young's modulus as a function of v_f depends only slightly on the properties of the polymer matrix.

Anti-adhesive treatment of the glass sphere surface has an influence on the measured E_c value, in particular the application of a liquid layer on the particle surface. In this case not only the adhesion but also the friction between the matrix and the particles is reduced. Theoretical computations and experimental results of Shaulov and Oshmian [157] in our laboratory demonstrate that in this case $E_c < E_m$, decreasing with the increase of filling (see Fig. II.18).

Several articles are related to the properties of oriented short fiber composites [159–161]. But the majority of reported experimental data were obtained on poorly characterized materials; in particular, the fiber-orientation distribution and the actual aspect ratio of the short fibers in the composite were not well defined.

Kacir et al. [162], thoroughly investigated the dependence of stress-strain properties on the fiber alignment distribution for brittle and ductile epoxy composites containing aligned short-glass-fiber bundles. In order to analyze the fiber-orientation distribution in the aligned, reinforced sheets, colored glass-fiber bundles were added to the fiberglass prior to the orientation. The angle of deviation of each colored bundle from the major alignment direction was measured and the bundle orientation distribution determined. Various other methods for the characterization of fiber alignment in opaque composites have been described, and recently reviewed by Kacir et al. [163].

The effect of the off-axis angle on the tensile strengths of aligned short-fiber composites has been treated theoretically by Azzi and Tsai [67] in the following way:

$$\frac{1}{\bar{\sigma}_x^2(\theta)} = \frac{\cos^4\theta}{\bar{\sigma}_x^2} + \left(\frac{1}{\bar{\tau}_{xy}^2} - \frac{1}{\bar{\sigma}_x^2}\right)\sin^2\theta\cos^2\theta + \frac{\sin^4\theta}{\bar{\sigma}_y^2} \tag{II.78}$$

where $\bar{\sigma}_x(\theta)$ is the off-axis tensile strength of the composite, $\bar{\sigma}_x$ is the tensile strength in the major fiber-alignment direction, $\bar{\sigma}_y$ the transverse tensile strength, $\bar{\tau}_{xy}$ is the in-plane shear strength along the main orthotropic axis, and θ is the angle between the direction of the applied external load and the major alignment direction.

The validity of this work has been confirmed by the measurement of the fiber-orientation distribution for three fiber bundle lengths at various angles [163].

An equation suggested by Leknitskü [73] has been used for the calculation of the tensile modulus $E_x(\theta)$:

$$\frac{1}{E_x(\theta)} = \frac{\cos^4(\theta)}{\bar{E}_x} + \left(\frac{1}{\bar{G}_{xy}} - \frac{2\mu_{xy}}{E_x}\right)\sin^2\theta\cos^2\theta + \frac{\sin^4\theta}{\bar{E}_y} \quad \text{(II.79)}$$

where $\bar{E}_x$, $\bar{E}_y$, and $\bar{G}_{xy}$ are the longitudinal, transverse, and in-plane shear moduli of the aligned fiber composites, respectively, and μ_{xy} is the major Poisson ratio.

The dependence of tensile modulus and tensile strength on the off-axis angle between major fiber alignment and tensile loading directions was in satisfactory agreement with the theoretical equations.

A number of other factors have to be considered when comparing the results of model calculations with experimental data:

1) Adhesion. Most models imply an absolute matrix-to-filler adhesion. This seems to be true for some real materials in the range of small deformations. However, one has to take into consideration compressive residual stresses arising due to a difference in the thermal expansion coefficients of filler and matrix at $T < T_g$ of the matrix. Moreover, systems may exist in which the adhesive bonds are distorted already in the initial stages of deformation. Usually, this is clearly indicated by an abrupt increase of the volume of a specimen under stress, corresponding to adhesive bond disruption, with the appearance of a pore at the matrix-filler interface.

2) Existence of pores in the initial material due to imperfection of its preparation technology. If the number of pores is known, it is possible to take into account their effect on the elastic modulus, Poisson's ratio, etc. assuming that one deals with a three-, rather than a two-component composite. Unfortunately, in most cases the material porosity can only be evaluated qualitatively, by comparing the calculated and measured densities of a specimen (or a product). The two factors (adhesion disruption and porosity) result in a decrease of elastic modulus and Poisson's ratio.

3) Nonuniform filler distribution: filler agglomeration. Most models imply a uniform filler distribution in the matrix, in the sense that the matrix layer around all particles is either equal or proportional to the particle size. At low filler concentrations, agglomeration does not practically affect the physico-mechanical properties; the elastic modulus is proportional to the bulk fraction of the filler. At higher filler concentrations, however, this dependence is nonlinear; agglomeration results in an apparent rise of the filler bulk fraction, i.e. the modulus is higher than expected from v_{fl}, due to the agglomeration. No quantitative analysis of this problem is available as yet.

4) Macroscopically nonuniform filler distribution in a specimen or in a product. This is a very important factor, which can cause the properties of different test samples to depend on the method of their preparation (e.g. the conditions of specimen injection molding); moreover, the properties of test samples and products may be different. This type of nonuniformity was taken into account by semiempirical calculations in terms of various "shell" models, which describe a specimen as consisting of a core and a shell with different filler contents [57, 58].

5) Distribution of fiber orientations in fiber filled composites. This distribution depends on the conditions of specimen preparation. As mentioned above, this factor makes

it difficult to analyze published data on short-fiber composites and to compare them with the idealized models described in Section 2.2. Besides, this factor casts doubt upon the representativity of especially prepared test samples.
6) Real shape of filler particles. The shape of disperse filler particles differs usually from an ideal sphere or rectangle. The particle shape effect on the properties of a material is generally assumed to be insignificant. As far as we know, however, an adequate quantitative analysis of this problem has not been made as yet.
7) Real distributions of disperse particle sizes and fibrous filler diameters. Real distributions always differ from an ideal uniform distribution, and also from the infinitely broad distribution which would be necessary to fill the entire space completely. It is commonly believed that polydisperse or three-phase models should be used for broad distributions, and regular models for narrow distributions. The difference between real and ideal distributions is especially large at high levels of filling (as the limiting filling level depends on the distribution type), and in the case of very stiff fillers. Additional theoretical quantitative analysis of this problem is also necessary.
8) Real fiber length distribution. A real fiber length distribution usually includes length above and below the critical value. This problem requires that the relevant expressions should be integrated, making allowance for the real distribution function. Such an approach involves numerous calculations and, certainly, necessitates the knowledge of the distribution function of fiber lengths.
9) Allowance for filler anisotropy. Some of the fibers used to reinforce composites, in particular synthetic polymer fibers, are essentially anisotropic by themselves. Evidently, this leads to significant changes of the elastic properties of a composite. Consequently, the calculations have to allow for filler anisotropy. In certain cases disperse fillers also show anisotropy, which should be borne in mind when comparing theoretical and experimental data.
10) Uncertainties in the polymer matrix properties. The physico-mechanical properties of polymer matrices depend on the degree of crystallinity, molecular weight, molecular weight distribution, etc. These parameters are functions of the processing conditions and may change after the introduction of a filler [59].

All these circumstances should be borne in mind when calculation results are to be compared with experimental data.

References

1. Dewey, J. M.: J. Appl. Phys. *18*, 578 (1947)
2. Decuntz, J. A., Hoffman S.: The Static Strength of Syntactic Foams, Trans. ASME, Ser. E., v. *36*, p. 551, 1965
3. Agarwal, B. D., Panizza, G. A., Brontman, L. J.: J. Amer. Ceram. Soc. *54*, 620 (1971)
4. Sahu, S., Brontman, L. J.: St. Louis Symp. Advan. Compos. 5th, April, 1971
5. Hashin, Z.: J. Appl. Mech., *29*, 143 (1962)
6. Kerner, E. H.: Proc. Phys. Soc. *69*, 808 (1956)
7. Van der Pool, C.: Reol. Acta *1*, 198 (1958)
8. Smith, J. C.: J. Res. Nat. Bur. Stand. A *79*, 419 (1975)
9. Christensen, R. M., Lo. K. H.: J. Mech. and Phys. Solids *27*, 315, (1979)
10. Zgaevskii, V. E.: Int. J. Pol. Mat. *6*, 109 (1977)
11. Manevich, L. I. et al: Hydroautomatics and Theory of Elasticity, Dnepropetrovsk (USSR) 1984, in press
12. Hershey, A. V.: S. Appl. Mech. *21*, 236 (1954)

13. Fujii, T., Zako, M.: Fracture of Composite materials, Japan (1978)
14. Michler, G., Gruber, K.: Plaste und Kautsch. *23*, 346 (1976)
15. Wagner, E. R., Robeson, L. M.: Rubb. Chem. Technol. *42*, 641, (1970)
16. Manevich, L. I. et. al.: Dokl. Akad. Nauk SSSR *270*, 806 (1983)
17. Knunyantch, N. N. et al.: Dokl. Akad. Nauk SSSR *270*, 803 (1983)
18. Givental, L. E. et al.: Thesis of Conference USSR composite materials, Tashkent, September 1983, YII, 118 (1983)
19. Gay, M. I. et al. ibid. YII, 77 (1983)
20. Batchelor, G. K., Groen J. T.: Fluid. Mech. *56*, 376 (1972)
21. McGee, S., McCullogh, R. L.: Polymer Compos. *2*, 149 (1981)
22. Hill, R.: J. Mech. Phys. Solids *13*, 213 (1965)
23. Budiansky, B.: ibid. *13*, 223 (1965)
24. Ashton, J. W., Halpin, S. G., Petit, P. H.: "Primer on Composite Materials: Analysis", Stanford Conn. Technomic, (1969)
25. Tsai, S. N.: Formulas for the Elastic Properties of Fiber-Reinforced Composites, AD 834851, National Technical Information Services, Springfield
26. Nielsen, L. E.: Appl. Polym. Symp. *12*, 249 (1969)
27. Lewis, T. B. Nielsen, L. E.: J. Appl. Polym. Sci. *41*, 26 (1970)
28. Progelhof R. C., Throne, J. L.: Polym. Eng. Sci. *19*, 493 (1979)
29. Richard T. G.: J. Comp. Mater. *9*, 108 (1975)
30. Kenyon, A. S., Duffey, H. J.: Polym. Eng. Sci. *7*, 189 (1976)
31. Smith, J. C.: J. Res. Nat. Bur. Stand. A80 (Phys. and Chem.) 45, (1976)
32. Isha, O., Cohen, L. I.: Int. J. Mech. Sci. *9*, 539 (1967)
33. Nielsen, L. E.: "Mechanical Properties of Polymer and Composites, v 2, New York Marcel Dekker 1974
34. Behrens, E.: J. Appl. Mech. *38*, 1062 (1971)
35. Chen, C. H., Cheng, S.: J. Appl. Mech. *37*, 186 (1970)
36. Hill, R.: J. Mech. Phys. Solids *12*, 199 (1964)
37. Hashin, Z.: AIAA Journal *4*, 1411 (1966)
38. Adams D. F., Doner, D. R.: J. Compos. Mater. *1*, 4 (1967)
39. Tsai, S. W., Adams, D. F., Doner D. R.: Analysis of Composite Structures, NASA CR-620, 1966
40. Vanin, G. A.: Dokl. Akad. Nauk USSR, A, N 4, 1976
41. Tarnopolskii, Yu. M. et al.: Mech. Polymer. USSR *4*, 676 (1971)
42. Kelly, A., Tyson, W. R.: High Strength Materials p. 578, New York: John Wiley 1965
43. Cox, H. L.: British J. Appl. Phys. *3*, 72 (1952)
44. Outwater, J. O.: Mod. Phys. Plast. *33*, 156 (1956)
45. Rosen, B. W.: Mechanics of Composite Strengthening Fiber Composite Materials, ASM 72, p. 75, 1965
46. Piggot, M. R.: Acta Met. *14*, 1429 (1966)
47. Russel, W. B.: Z. Angew. Math. Phys. *24*, 581 (1973)
48. Krenchel, H.: Fiber Reinforcement Copenhagen: Akademisk Forlag, 1964.
49. Lees, K.: Polym. Eng. Sci. *8*, 195 (1968)
50. Chen, P. E.: ibid. *11*, 51 (1971)
51. Christensen, R. M.: Mechanics of Composite Materials, New York; John Wiley, 1979
52. Sendenckyi, G. P.: in Composite Materials, v. 2, New York: Academic Press, 1974
53. Christensen, R. M.: J. Eng. Mater. Techn., 101, (1979)
54. Padawer, G. E., Beecher, N.: Polymer. Eng. Sci. *10* (3), 185 (1970)
55. Riley, V. R.: Interaction effects in fiber composites. Polym. Conf. Series, Univ. of Utahs, June 1970
56. Piggot, M. R., Glavinchevski, P.: S, Mater. Sci. *8*, 1373, (1973)
57. Okagawa, A., Mason, S. G.: Can. J. Chem., *53*, 2689 (1975)
58. Utracki, L. A., Fisa, B.: Polym. Compos. *3*, 193 (1982)
59. Curtis, A. C., Hope, P. S., Ward, I. M. ibid. *3*, 138 (1982)
60. Hashin, Z., Shtrikman, S.: J. Appl. Phys. *33*, 3125 (1962)
61. Kerner, E. H.: Proc. Phys. Soc. *B69*, 802 (1956)
62. Hashin, Z.: Theory of fiber-reinforced materials, NASA, CR-1974, (1972)

63. Brydges, W. T., Gulati, S. T., Baum, G.: J. Mat. Sci. *10*, 2044 (1975)
64. Levin, V. M.: Izv. ANSSSR, MTT, 88, (1967)
65. Greszczuk, L. B.: Soc. Plastics Ind., 20th Ann. Meeting of R.I.P. Div. Section 5C, p. 10
66. Tsai, S. W.: in Strength Theories of Filamentary Structures, Fundamental Aspects of Fiber Reinforced Plastic Composites, R. T. Schwartz, H. S. Schwartz (ed.) New York: John Willey, 1968
67. Azzi, V. D., Tsai, S. W.: Exp. Mech. 283, September (1965)
68. Fischer, L.: J. Engng. Ind., August (1967)
69. Petit, P. H., Waddoups, M. E.: J. Compos. Mater *3*, 2 (1969)
70. Golden blat, I. I., Kopnov, V. A.: Mechan. Polym. (USSR) *1*, 70 (1965)
71. Ashkenazi, E. K.: ibid. *2*, 79 (1965)
72. Tsai, S. W., Wu, E. M.: J. Compos. Mater. *5*, 58 (1971)
73. Lehnitzkii, S. G.: Theory of elasticity of anisotropic solid, Moscow: Nauka 1977
74. Annin, B. D., Baeyev, L. V.: in Fracture of composite materials, Riga: Zintne, p. 167, 1979
75. Goldenblat, I. I., Kopnov, V. A.: Criteria of Strength and ductility of constructional materials, Moscow: Mashinostroenie 1968
76. Skudra A. M., Bulavs, Ph. Yu., Rotzens, K. A.: Creep and static fatigue of Reinforced Plastics, Riga; Zinatne 1971
77. Rowlands, R. E.: AMD-v 13, Inelastic Behavior of Composite Materials, New York, 1975
78. Van Pho Phi, G. A.: Theory of Reinforced Materials, Kiev: Naukova Dumka 1971
79. Wu, E. M.: Phenomenological anisotropic failure criterion — In Composite Materials v. 2, G. P. Sendeckyj (ed.), New York: Academic Press 1974
80. Vicario, A. A., Toland, R.: In Composite Materials Brontman, L. J., Krock (ed.), v. 7, New York, Academic Press, 1975
81. Argon, A.: Composite materials, v. 5: Fracture and Fatigue, New York Academic Press, 1974, p. 121
82. Hedgepeth, I., Van Dyke, P. J. Comp. Mater. *1*, 294 (1967)
83. Franklin, G.: Fiber Sci. Techn. *2*, 241 (1970)
84. Wright, P., Ebert, L.: Met. Trans. *3*, 1645 (1972)
85. Zweben, C.: J. Mech. Phys. Solids *22*, 193 (1974)
86. Kuper, D., Piggot, M. R.: Fracture, v. 1, New York, Pergamon Press, 1978
87. Lexovskii, A. M. et al.: in Physic of Strength for Composite Materials, Leningrad: 1979, p. 256
88. Hendrepeth, J. M.: NASA, Techn. Rep. TN-D 882 (1961)
89. Herring, H., Lytton, J., Steele, I.: Met. Trans., *4*, 807 (1973)
90. Weibull, W.: Ing. Vetensk. Akad. Handl. 151 (1939)
91. Weibull, W.: ibid. 153 (1939)
92. Coleman, B. D.: J. Mech. Phys. Solids, *7*, 60 (1958)
93. Daniels, N. W.: Proc. Roy. Soc. (London) Ser. A, *183*, 405 (1945)
94. Gucer, D. E., Gurland, I. J.: Mechan. Phys. Solids *10*, 365 (1962)
95. Rosen, B. W.: AIAA. J. *2* (11), 1985 (1964)
96. Ovchinskii, A. S., Nemtzova, S. A., Kop'ev, I. M.: Mechan, Polymer. 800 (1976)
97. Kop'ev, I. M., Ovchinskii, A. S., Bilsagaev, N. K.: Fracture of Composite Materials, Riga, Zinatne 1979, p. 57
98. Kompaneetz, L. V. et al.: Dokl. AN SSSR 264, 1425 (1982)
99. Schmitz, G. K., Metcalfe, A. G.: Effect of Length on the Strength of Glass Fibres, Preprint N 87, ASTM, Philadelphia, Pensilvania, 1964
100. Turusov, R. A. et al.: in Trudi CNIISK, N 53, Moscow. Stroyizdat 1975, p. 72
101. Berlin, A. A.: Thes. of I-st Allunion Sympos. Compos. Materials, Tashkent: Sept. 1980
102. Mileiko, S. T.: J. Mater. Sci. *4*, 874 (1969)
103. Protasov, V. D.: Mechanic of Compos. Mater., N 5, 835 (1980)
104. Zelenskii, E. S. et al.: Dokl. AN SSSR, *260*, 1099 (1981)
105. Gunyaev, G. M.: Structure and Properties of Polymer Fibre-Reinforced Composites, Moscow: Chimija, 1981
106. Rabinovich, A. L.: Introduction in Mechanic of Reinforced Plastics, Moscow: Nauka 1970, p. 482

107. Skudra, A. M., Bulavs, Ph. Ya.: Structural Theory of Reinforced Plastics, Riga, Zinatne 1978, p. 192
108. Inglis, C. E.: Inst. Naval Archetets, Trans. 55, 219 (1913)
109. Griffith, A. A.: Phil. Trans. Roy. Soc. (London) 221 A, 163 (1920)
110. McClintock, F. A.: Symp. High Performance Compos. 4th, St. Louis, Mo 1969
111. Bazhenov, S. L. et al.: Dokl. AN SSSR *277*, 854 (1981)
112. Cooper G. A., Kelly, A. J. Mech. Phys. Solids 15, 279 (1967)
113. Cooper, G. A.: Composite materials, V. S., Fracture and Fatigue, New York, Academic Press, 1974,
114. Guz, A. N.: Appl. Mechan. *18*, N9, 3 (1982)
115. Bolotin, V. V., Bolotina, K. S.: in Strength of Materials and Constructions, Kiev; Naukova Dumka 1975, p. 231
116. Korotkov, V. N.: Mechan. Compos. Mater. N 2, 290 (1983)
117. Vinogradov, V. M.: in Plastics for Constructive Use, Moscow: Chimija 1974, p. 46
118. Bolotin, V. V., Vorontzov, A. N.: Mechan. Polymer. N 5, 790 (1976)
119. Afanasev, Yu. A. et al.: Mechan, Compos. Mater. N 4, 651 (1980)
120. Adams, D. F., Doner, D. R.: J. Compos. Mater., *1*, 152 (1967)
121. Foye, R. L.: Proceedings of the 10th National SAMPE Symposium G-31 Nov. 1966.
122. Foye, R. L.: Inelastic Behavior of Composite Materials, v. 13, New York. 1975,
123. Chamis, C. C.: Composite Materials, v. 5, Fracture and Fatigue, New York, Academic press, 1974, p. 73
124. Foye, R. L.: Compression Strength of Unidirectional Composites, AIAA Paper N 66-143, AIAA 3rd Aerospace Science Meeting, New York, 1966
125. Melbardis, Yu. G., Crechers, A. V.: Mechan. Compos. Mater. N 4, 601 (1982)
126. Golovkin, G. S.: Plast. Massy, N 6, 28 (1981)
127. Perov, B. V. et al.: Fracture of Composite Materials, Riga, Zinatne 1979, p. 182
128. Kulkarni, S. V. Rice, J. S., Rosen, B. W.: NASA CR-112334, 1973
129. Kuperman, A. M. et al.: Thes. of 3rd Allunion Symp. Phys. of Strength of Compos. Materials, Leningrad: LIYaF, 1979, p. 50
130. Bennett, S. S., DeVries, K. L., Williams, M. L.: Int. J. Fracture *10*, 33 (1974)
131. Manevich, L. I., Pavlenko, A. V.: J. Appl. Mech. and Tech. Phys. USSR-, 1982, N 3, 140
132. Dow, N. F., Grunfest, I. S.: General Electric, TIS 60 D389, 1960
133. Rosen, B. W.: Mechanics of Composite, Strengthening, Fiber Composite Materials — ASM72, 1965, p. 75
134. Hayashi, T.: AIAA Paper N 65–770, 1965
135. Schuerch, N.: AIAA Journal, *4*, 102 (1966)
136. Greszczuk, L. B.: ASTM STP N-546, Philadelphia, Pa, 1974
137. Rabinovich, A. L.: in Trudi MPhTI, N 7, Moscow 1961, p. 49
138. Landau, L. D., Lifshitz, E. M.: Teoriya uprugosti, Moscow, Nauka, 1965, p. 204
139. Levenetz, B.: Proc. 19th Conf. SPI Reinforced Plastics Div. 1964, Sec. 14D
140. Levenetz, B.: Proc. 20th Conf. SPI Reinforced Plastics Div. 1965, Sec. SF
141. Natrusov, V. I.: Technology, physico-technical properties and use of Glass-reinforced plastics, Moscow; VNIISPV 1975, p. 92
142. Roginskii, S. L., Kanovich, M. Z., Koltunov, M. A.: High Performance Glass-Fiber Reinforced Plastics, Moscow: Chimija 1979, p. 22
143. Kulkarni, S. V., Rice, G. F., Rosen, B. W.: Composites, *6*, 193 (1975)
144. Kalnin, I. L.: in Fracture of Composite Materials, Riga: Zinatne 1979, p. 221
145. Shyne, J. I., Milewski, V.: Proc. Ann. Techn. Conf. of RP/Composites Inst. Soc. Plast. Ind., v. 24, 1969, 18D
146. Kobetz, L. P. et al.: Mechan, Polymer., N 4, 579 (1978)
147. Greszczuk, L. B.: AFML-TR-72-107, Wright Patterson AFB, Ohio 1972
148. Do'akova, V., Inacek, J.: Faserborsch. und Textiltechn. 57, *29* (1978)
149. Tovmasyan, Yu. M., Topolkaraev, V. A. et al.: DAN USSR *270*, 649 (1983)
150. Iisaka, K., Shibayama, K.: J. Appl. Polym. Sci. *22*, 1321 (1978)
151. Trachte, K., Benedetto Di, A. T.: Int. J. Polym. Sci. *1*, 75 (1971)
152. Schmidt, L. R.: Polym. Eng. Sci. *17*, 666 (1977)
153. Narkis, M., Joseph, E.: J. Appl. Polym. Sci. *22*, 3531 (1978)

154. Yilmazer, U., Farris, R. J.: Polym. Compos. *4* (1), 1 (1983)
155. Friedrich, K., Karsch, V. A.: ibid *3* (2) 65 (1982)
156. Tanaka, K., Kobunshi Roubunshu *37*, 575 (1980) *38*, 11 (1981) *39*, 79 (1982)
157. Shaulov, A., Oshmian, V., Enikolopian, N. S.: unpublished results
158. Kalinski, R., Galeski, A., Kryszewski, M.: J. Appl. Polym. Sci., *26*, 4047 (1981)
159. Kelly, A., Tyson, W. R.: J. Mech. Phys. Solids, *13*, 329 (1965)
160. Rosen, W.: in Fiber Composite Materials, American Society for Metals, Metals Park, Ohio (1965)
161. Blumentritt, B. F., Vu, B. T., Cooper, S. L.: Composites, *6*, 105 (1975)
162. Masoumy, E., Kacir, L., Kardos, J. L.: Composites, *4*, 64 (1983)
163. Kacir, L., Narkis, M., Ishai, O.: Polym. Eng. Sci., *17*, 234 (1977)

3. Polymer Composites: Manufacturing Principles

The history of filled polymers began at the turn of this century when numerous empirical experiments led to the development of a range of technically valuable materials on the basis of phenol- and urea-formaldehyde resins. Wood flour, linter and asbestos, paper and cloth were the first fillers to be used industrially. If we regard this early work from the point of view of contemporary knowledge, we see that all basic filler types — inorganic, organic, synthetic, dispersed, fibrous and lamellar — had already been embraced at that early stage.

The filling of thermosetting plastics may rightfully be called a traditional manufacturing operation. The low viscosity of the starting oligomers made the task of impregnation and uniform distribution of the filler a very easy one. The adhesion of most thermosetting plastics to most fillers is excellent; hence the problem of interphase contact between filler and matrix was not acute.

A radical change came in the early 40's, when the production of glass fibers and glass fiber-reinforced materials begun in USA. The manufacture of articles reinforced with continuous fibers called for a new winding technology. To enable the use of glass mats as reinforcement, the so-called sheet molding compound (SMC) and bulk molding compound (BMC) techniques were elaborated.

Such "old" matrices as phenolic and aminoformaldehyde resins turned out to be suitable for glass fiber reinforced materials. But novel resins, epoxy and especially polyester, have also been employed on a large scale.

By the beginning of the 70's, the situation on the world plastics market was quite peculiar. Almost all thermosetting resins on the market were filled structural or coating materials, as well as adhesives. At the same time, virtually all thermoplastics were unfilled. Since the product output was 3 to 1 in favor of thermoplastics, the unfilled plastics were clearly dominant at that time.

Poly(vinyl chloride) (PVC) was apparently the first thermoplastic to be systematically filled with mineral materials. Linoleum, decorative tiles, artificial leather are examples of products where only the combination of a plastic and a mineral filler (chalk, clay) could produce the desired consumer effect.

The thermoplastics marketing boom falls in the beginning of the 70's. Following the assessment of other authors, there are several reasons for it. Oil crisis and hopes for production cost cuts owing to the use of cheaper mineral fillers served as baits. The aftermath of the oil crisis is still perceptible today, to a greater or lesser degree. As far as the hopes for cheaper materials, retaining all the inherent properties of the plastics, they turned out to be ephemeral for most thermoplastic materials.

Very soon, however, other problems came to the fore, viz. upgrading of mechanical and technological properties, flame retardancy, or imparting som special functional properties. It is known that most companies engaged in the production of polymers have been carrying out extensive investigations into the perspectives opened up by the blending technology, hoping to come to an optimum performance vs. economy tradeoff.

Now that a decade has elapsed since the first boom, we may draw some conclusions. The filling of thermoplastics did not lead to any principally novel manufacturing processes. All producers stuck to the conventional powder and melt compounding processes, using available mixing machines.

The approach to article manufacturing was just as traditional. Injection molding, extrusion and thermal molding have been found quite suitable for processing particle-filled thermoplastics into articles.

The "attack" of fillers on the commodity plastics did not come out to be a major success. They did invade, however, such fields as the production of cable sheaths from PVC and drain pipes from rigid PVC and polyethylene. The output of dispersion-filled polypropylene amounts to several dozens of thousands of tons per year, which is impressive by itself but, in fact, does not reach to even 1% of the total output of this plastic. The filling of polyethylene and polystyrene is done usually in pursuit of very special purposes.

The situation on the engineering plastics market appears to be much more fortunate. Most are now sold filled, the main filler being short glass fibers. Compared to unfilled plastics, this new generation of structural materials features higher mechanical strength and rigidity, lower thermal deformation and a better dimensional stability in machining. But often these achievements come at the expense of higher brittleness, poorer surface condition and higher wear of the working parts of machines.

The requirements to the microstructure of polymer composites become more stringent with the progress in understanding the physical chemistry of filled systems and the development of structure analysis methods. Below we give the main parameters of the individual components, and the composite system as a whole, that must be controlled during the manufacturing process.

Filler:

1. Filler particle structure and shape stability in the composite, including
   a) to keep the reinforcement fiber length above the critical value, to control size distribution;
   b) to keep the characteristic aspect ratio constant for lamellar fillers;
   c) to control dispersed filler particle agglomeration and aggregation;
2. Preservation (or modification) of the chemical properties of the filler particle surface;
3. Control of the content of absorbed moisture and adsorbed gases on the filler particles.
4. Uniformity and completeness of filler particle/coupling agent interaction.*

* A coupling agent is a chemical providing strong links between the filler particles and the matrix; substances such as silanes, organic titanates, stearates, chlorinated paraffins, oligomers, etc. are used for this purpose.

Matrix:

1. Modification of the molecular characteristics of the matrix during mixing with the filler;
   a) variation of the molecular weight distribution and the average molecular weight;
   b) variation of the degree of branching, formation of a crosslinked microgel;
2. Variation of structural parameters: degree of crystallinity, T_g;
3. Variation of the chemical structure of the matrix on account of filler-catalyzed processes.

Composite:

1. Distribution of the filler particles in the bulk of the material
   a) uniformity of distribution,
   b) degree of orientation of the reinforcing filler particles;
2. Pore ratio
3. Distribution of coupling and modifying agents;
4. Stability of filler/matrix interaction.
5. Interface characteristics.

In cases where the filler itself is a composite, e.g. a gas-filled material, or where a material with a predetermined structure is to be obtained, the number of parameters to be controlled increases correspondingly.

We hope that we have made it clear in the two preceding chapters that adequate methods of composite structure analysis are either not available, or cannot be used commercially because of their high complexity and cost. Therefore the parameters listed above are largely speculative. But we will call them to mind every time we will compare different manufacturing processes.

Blending of the filler with a polymer, monomer or oligomer is the key step of composite production. Methods aiming at improving the phase interaction conditions may be classified as follows:

Filler treatment:
Heat treatment
Physical treatment
Treatment with coupling agents
Chemical treatment

Filler/polymer contact:
Blending in molten plastic
Coating filler particles with a polymer
Monomer polymerization on the filler surface
Grafting of a polymer to the filler surface

Polymer matrix modification:
Mixing of different polymers
Amorphization of initially crystalline polymers
Addition of functional groups
Copolymerization
Crosslinking of polymers.

Taking this classification as a basis, we will now consider the main technological trends.

3.1 Compounding

The preliminary step is to mix a dispersed filler with a powdered or granulated polymer in a high-speed dry mixer. Powdered polymers may, obviously, provide a better homogeneity of the final product. Besides, it is economically unfeasible to granulate the polymer unless it has been produced by bulk-polymerization (e.g. LDPE, impact polystyrene).

Adsorbed gas may be liberated at this first mixing step; in some processes also vacuum, heating, or blowing of the filler with inert gases is used.

The central step of thermoplast compounding is blending the filler with molten polymer.

3.1.1 Rheology of Filled Thermoplastics

Compounding of molten thermoplastics can be described quantitively by rheological equations. Molten polymers are non-newtonian liquids and have the characteristic viscosity anomaly under the effect of shear deformation. The rheology of melts filled with rigid particles is determined by the dispersion structure, the shape and orientation of the dispersed particles, and by the interaction between particles. Much work has been devoted to polymer rheology, but the rheological behavior of filled systems has not received comparable attention so far. These problems are dealt with in reviews by Van Oene [1], Vinogradov and Malkin [2] and, more recently, Utracky and Fisa [175].

One of the remarkable features of the behaviour of dispersed solid particles in viscoelastic liquids is their migration toward the center of flow, and the tendency to form a "nut-in-a-shell" structure. This effect causes depletion of the surface layer and concentration of the filler at the center of flow. In capillary flow this causes a considerable reduction of dispersion viscosity [3].

Interactions between particles depend on their shape, size, structure and surface composition and may have a variety of effects on the rheological behavior, including even an increase of viscosity with increased shear stress. It follows from this that most real systems have to be subjected to experimental investigations (see, e.g. [4,5]).

An attempt has been made to find a unifying approach to flow curves of thermoplastics at different temperatures, independent of filler parameters [6]. To this end, the rheological functions of LDPE, HDPE, PP, PS, nylon, polyester and polycarbonate, as well as of 12 types of fillers, and 7 types of surface modifiers were investigated. Another approach uses master curves, i.e. curves invariant with respect to filler parameters (type, quantity, size, shape) [7].

Deryagin et al. [176] have shown that the interaction between small particles may follow several different mechanisms. According to data published by Plueddemann [177], the filler-filler interaction is stronger than the filler-polymer interaction. The role of many coupling agents consists, apparently, in a reduction of the filler-filler interaction. It has been reported [178] that, after treatment with a coupling agent, the viscosity of a suspension may become even smaller than that of the polymer melt.

Different equations describing the $\eta = f(v_f)$ relationship, the effect of polydispersity and filler particle shape, have been analyzed in a recent review article [175]. The authors have considered the effects due to shear deformations in non-Newtonian liquids, such as yield stress, thixotropy, rheopexy (anti-thixotropy), dilatancy and pseudoplasticity. The data of a total of 36 most recent experimental works dealing with commodity thermoplastics filled with glass fibers, chalk, limestone, talk, TiO_2, carbon black, mica, $CaCO_3$, and glass beads are listed in tables. The autors give a number of practical hints. One example is that in high-filled systems the time dependence of viscosity is so tangible that the authors recommend to mold them immediately after compounding.

3.1.2 Blending Effectiveness Criteria

A numerical estimate of the mixing quality is based on a statistical approach. In an ideal binary mixture, the scatter of concentrations of the dispersed substance follows the binomial law. The probability $P(b)$ for b particles of the dispersed phase to be present in an analyzed sample of the composite is equal to the binomial distribution density

$$P(b) = \frac{(n/q)!}{b!(n/q - b)!} q^b (1 - q)^{n/q - b},$$

where n is the average number of particles in the composite, b is the number of particles in the analyzed sample, q is the volume concentration of the dispersed particles.

The variance of this theoretical distribution (ξ^2) is:

$$\xi^2 = n(1 - q)$$

A mixing index I is defined as the ratio of the theoretical to the experimental variance (S^2):

$$I = \frac{\xi^2}{S^2}$$

The magnitude of the mixing index is a function of the shear deformation and, up to a certain value of the latter, also depends on the type of mixing apparatus. An ideal mixture with $I = 1.0$ is practically unrealizable even after very long mixing times.

The overall compounding process includes the steps of homogenization, distribution and dispersion. Homogenization and distribution of different components (fillers, pigments) in a thermoplastic is effected in a laminar flow of the melt (low Reynolds number), under large shear stresses. This step requires extended time. Dispersion, or comminution of the agglomerates of the dispersed phase, occurs under the effect of the large shear stresses arising due to the relative motion of the viscous polymer and the solid particles. To break up an agglomerate, the viscous friction forces on its surface must be greater than the strength of interaction between individual particles. An effective deagglomeration consists in separating the particles to distances longer than the radius of action of the cohesion forces.

It is known that for each system there is a critical shear stress below which no dispersion can occur. The degree of dispersion is controlled by the work of dispersion and does not depend on temperature nor on motion velocity. However, in a low-viscosity dispersion medium (e.g. at high temperature) the shear stresses are low and large shear deformations are needed to produce the effect.

In Section 1.3.1 we considered an alternative approach to the blending efficiency problem. It was suggested there that the spatial distribution of filler particles may be estimated by a function of the momentum of filler agglomerates. This approach does seem to be a very promising one for practical applications.

3.1.3 Physico-Chemical Effects Involved in Compounding in a Melt

Compounding can be carried out in machines of different design. Mechanical devices based on the shear deformation principle have become most popular. The family of tools consists of rollers, rotors, and screw mixers. The treatment of suspensions with ultrasound [8], electrohydraulic shock, and other techniques are also of a certain interest. In any case the mechanical impact exerted on the melt results in a non-uniform distribution of internal stresses. The critical stresses developed at certain segments of macromolecular chains may cause the breaking of covalent bonds and the formation of macroradicals. Depending on the nature of the polymer and the processing conditions, the secondary reactions may take different routes. Such effects as variation of the MWD, decrease or increase of the viscosity average molecular weight, and crosslinking of macromolecules accompany the processing of unfilled polymers [9], and must also be taken into account in processing filled polymers.

For example, in the course of processing a chalk-filled LDPE one can observe either a decrease of the viscosity average molecular weight (at high melt temperatures), or an increase (at low temperatures), which is due to the different conditions under which degradation and crosslinking processes occur [10]. A similar effect was observed for chalk-filled PVC [11].

3.1.4 Extrusion Compounding

Mixing screw extruders have become the most popular tools for compounding disperse or short-fiber fillers with thermoplastics, mainly due to their economic and continuous operation. All operate with laminar blending of the components, though they have different efficiency from one design to another.

Single screw extruders, with a screw length to diameter ratio of more than 20:1, find application mainly in the United States for blending chopped glass fibers with thermoplastic materials. The problem with the blending of this filler is that in the charging zone it must have a bulk mass whereas in the melt is must be distributed into separate filaments. Single screw extruders are always used in combination with high-speed dry-effect mixers in the same production line. Yet, the blending efficiency attained on such lines is not always high enough.

Much more efficient (though much more expensive) are "Co-Kneters", or oscillation mixers, manufactured by, e.g., BUSS, Switzerland. The screw of these machines carries out rotary and translatory motions. The screw thread is intermittent, and special kneading teeth secured on the apparatus body are adapted to enter the grooves between the thread segments. The variable screw motion assures an intensive displace-

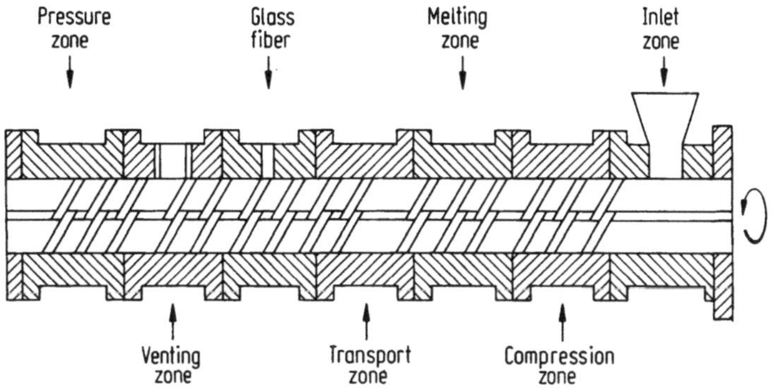

Fig. III.1. Schematic view of a double-screw extruder for glass fiber filling of thermoplastics.

ment of material and thereby a uniform blending of the components over a relatively short screw length (the L/d ratio of such a Co-Kneter is (7–12):1).

Two-screw extruders have become widerspread in Western Europe and USSR. The major manufacturers of this kind of machines are the West German firms WERNER & PFLEIDERER, HERMANN BERSTORFF, and KRAUSS MAFFEI. They permit a fine control of blending conditions, temperature, and permit the operations to be carried out in different reaction zones.

The schematic view of a double-screw extruder is shown in Fig. III.1. The blending efficiency is improved owing to the provision of two mutually perpendicular flows in the screw channels. Intensive shear deformations are produced in the clearances between the crews and also due to the provision of special kneading disks, gears, and reverse-thread screw sections.

The very complicated pattern of melt motion in the working zone of a double-screw extruder makes its theoretical analysis difficult, if at all possible. This is probably the reason for the diversity of extruder design. Besides, each machine offers a high potential for modifications in screw shapes, addition of extra elements, since in most cases the screws are assembled from sections.

A double-screw extruder can assure high productivity when working with a powdered polymer. The filler may be added in the charging zone (generally after preliminary mixing with polymer in a dry mixer). Other agents such as stabilizers, inhibitors, pigments, lubricants, etc. are added together with the filler.

In the case of glass fibers, glass powder, and other abrasive fillers, the manufacturers recommend adding of the fillers into the molten polymer in order to prevent rapid wear of the moving parts in the charging zone. When filler concentrations are to be high, it is best to feed some of the filler with the polymer and add the rest to the melt.

Volumetric or gravimetric batching devices are used for charging the components. Volumetric devices assure a ± 2–4% accuracy; the gravimetric method is more accurate (within ± 0.25–1%) but also more complex. The main problem of filler batching is related to filler fluidity. It is well known, for instance, how difficult it is to meter glass fibers. Despite the fact that milled short fibers are available, many manufacturers prefer charging the filler directly into the melt, with the fibers being milled by special elements secured onto the screw shafts.

The feeding of fillers with particle size >50 μm meets with no difficulties and can be done using any type of batching device (bucket wheel, shaker device, screw, conveyor, etc.). But talc, graphite, carbon black, kaolin, and other fillers characterized by a particle size of less than 10 μm have a poor looseness, the particles tend to adhere.

Since the feed zone design is important for the achievement of a given throughput efficiency, extruder designers have been paying special attention to the kneading disk and screw profiles, ever since the early patents of Erdmenger [12, 13].

The internal friction of the material under the effect of shear deformations produces enough heat to account for 70–80% of the total heat needed to melt the polymer in the melting zone. The rest is transferred through the walls from heating elements.

Sometimes, to accelerate melting, the modern double-screw extruders contain special meshing kneading disks which increase the shear effect. Special disk designs have been patented by manufacturers [14].

A dispersion zone may be located subsequent to the melting zone; it serves to disperse and distribute the filler particles in the molten polymer. For a better mixing effect in the laminar flow, some extruders are provided with a set of toothed mixing rings. Passing through them, the flow of molten material is divided into a plurality of layers, the number of which is a function of the number of teeth on the ring and the number of ring pairs. The number of layers may be as high as 10^9, indicating the excellent mixing effect achieved by this method. Mixing rings assure blending of material layers without considerable shear deformations. This is especially important for glass fiber-filled composites where excessive destruction of fibers must be prevented. Nevertheless, practical experience shows (Fig. III.2) that the double-screw extruder hardly permits the glass fiber length in the composites to be more than 0.35 mm.

Since it is virtually impossible to preclude large amounts of air adsorbed on the powdered material from getting into the melt, practically all modern double-screw extruder models are provided with a venting zone.

A considerable pressure is developed in the screw channel in order to overcome the

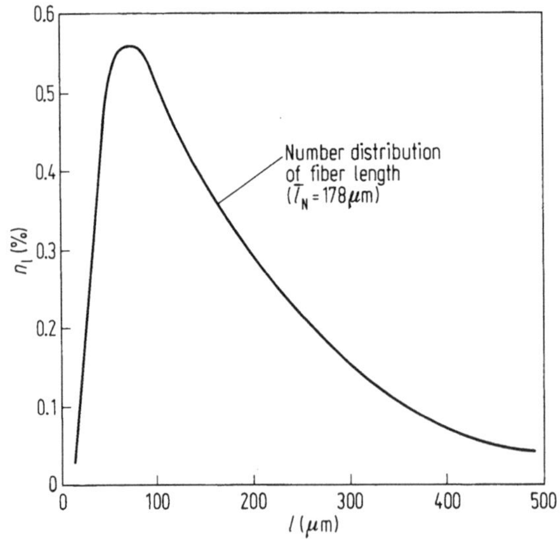

Fig. III.2. The number distribution of glass fiber length in a poly(butylene terephthalate) (PBT) composite after double-screw extrusion. $v_f = 0.17$. The starting fiber is continuous. (Fiber lengths measured microscopically.)

flow resistance. At the same time, energy dissipation from the melting zone must be minimized, and temperature peaks are undesirable. Designers have to seek a trade-off to find a satisfactory solution. The usual screw length/diameter ratio is from 20:1 to 30:1. The effects of screw configuration and compounding technique on the mechanical properties of filled thermoplastics are discussed by several authors [15–17].

Optimization of the mixing conditions in an extruder may be carried out using a mathematical planning model. It has been shown for glass-filled acetal resin and poly(butylene therephthalate) that five variables may be optimized within a total of eight experiments [18]. The effect of the MWD of the matrix has also been taken into account in that work.

To granulate polymers filled with rigid fillers, the knives and orifices must by made wear-resistant. But even then they will be worn out much sooner than in case of unfilled polymers.

3.1.5 Methods of Filler Treatment

Frissel [19] has suggested the following classification of polymer-filler interactions:

1. A simple mixture of a filler with a nonpolar polymer results in simple dilution and reduction of strength.
2. Wetting of the filler surface by the polymer, and good physical contact between the phases improves strength.
3. Chemical bonds across the interface is the ideal to be aspired for.

Most technologists agree that a good contact between filler and polymer is a positive factor. The simplest case is compounding of a filler with a polar polymer such as PVC. The thermodynamics of this process has been reported by Tager et al. [20,21]. Donor-acceptor interactions of the acid-base type are responsible for the effective contact. The interaction can be estimated by different methods, viz. from the temperature dependence of adsorption isotherms, calorimetrically, or by IR spectroscopy [22].

A fresh-formed surface of solid minerals has a high surface energy. Many techniques have been proposed in the patent literature [23] for a combination of mineral milling in ball or jet mills and compounding with polymers. Yet, the efficiency of such physical activation methods is dubious and the inaccuracy of the results is high. It seems likely that the leading role in the observed effects is played by the removal of surface moisture due to a local temperature rise under impact.

Coupling agents are commonly recognized as the best means of improving the filler-polymer interaction. A detailed analysis of the mechanisms by which these agents interact with the filler and the polymer matrix is often very complex. Fatty acids and their salts are traditionally added to certain thermoplastics to improve their processing properties. They have been found effective also in chalk-filled PVC compositions as they decrease the melt viscosity and thereby improve the technological fluidity. With the same purpose wax and low-molecular polyolefins are added to chalk and kaolin-filled PVC compositions. It has been recommended [23] that the melt temperature of the additives should be slightly lower than that of the matrix in order to promote filler dispersion with a high efficiency. An excessive amount of such additives is detrimental because, having a low viscosity, they decrease the shear deformations during the kneading operation.

Table 3.1. Progress of efficiency of coupling agents for glass reinforcement in polyesters [24)]

Date	Coupling Agent	Flexural strength (MPa)	
		dry	2 h water boil
1948	None	386	234
1950	BJY finish*	441	386
1960	Methacrylatochrome complex	503	428
1960	Vinyl silane	462	414
1962	Methacrylate silane	620	586
1972	Cationic vinylbenzyl silane	634	566

* Equimolar $CH_2{=}CHSi\,Cl_2$ and $CH_2{=}CCl{-}CH_2OH$

Silane derivatives and organic titanates are the most universal modern coupling agents. Originally they were developed for glass-filled polymers, especially unsaturated polyesters. Plueddeman [24)] quotes impressive figures on how the flexural strength of glass fiber reinforced polyesters has been improved in 24 years owing to the use of new, more effective coupling agents (see Table 3.1).

The activity mechanism of silane-based coupling agents has been discussed in many papers [25, 26)]. The simplest model of their behavior assumes the formation of covalent bonds with the resin during the curing, and oxone bonds with minerals. The oxone bonds such as

$$R{-}Si{-}O{-}Me$$

are of ionic nature and their stability versus hydrolysis is determined by the equilibrium constants. It is beyond the scope of this book to go into the details of the mechanisms of action of the various coupling agents, but it needs to be stressed that the real situation is far from being that simple. Plueddemann [26)] has named five factors to which the silane effect may be attributable: improved filler-matrix adhesion, protection of the filler surfaces from microflaws which initiate failure, reinforcement of the interface layer, improved filler wetting and dispersion conditions, increased hydrophobicity of the surface. The concept of a silane "monolayer" had to be dropped, and was replaced by the theory of structural gradients in the layers [27)]. The authors of the most recent works on silane coupling agents (Ishida [27)], Plueddemann [26)]) stress the polymeric structure of the interfacial layer which includes an interpenetrating network of polymeric siloxane segments and the matrix polymer, without any chemical cross-linking. This is assumed to be the decisive difference between low-molecular weight and high-molecular weight coupling agents.

There are different methods for adding silanes to the matrix, including dry mixing at room or elevated temperature, dispersion in water, dissolution in alcohols or organic solvents. For each particular composite, a specific silane coupling agent must be selected from the marketed assortment and use in the specified concentration.

But while the effect of these agents is quite apparent in glass-fiber composites, the situation is not as straightforward for disperse fillers. Admittedly, the agents improve the detachability of these fillers and reduce the effective viscosity of composites based on thermoplastic materials [28]. However, as regards the effect of silane treatment on the mechanical properties of, e.g., filled polyolefins, the reported data have been contradictory [29, 30].

Plueddemann [24] has described a test (Daniel flow point test) which allows to estimate the effect of the treatment of disperse fillers with silane.

Organic titanates are competitive with silane coupling agents. The commercial output of organic titanates has been increasing at a very fast rate since 1978 and by now has caught up with the production of silanes [31, 32].

The effectivity of these two basic types of coupling agents is corroborated by the fact that in 1981 as many as 15 US producers of disperse fillers were selling their products already treated with one of these coupling agent [33]. Although the addition of coupling agents such as silanes or organic titanates in amounts of 0.1 to 2.0% involves some extra cost and is not always well-argumented scientifically, the boom in their production is continuing. Specific coupling agents are on the market for many plastic materials including heat resistant polymers.

The effect attainable by chemical modification of the filler surface is closely related to the nature of the surface. One factor common in many types of fillers is the presence of hydroxyl groups on their surface. The hydrophilic properties of the surface are also determined by the presence or absence of acid groups. Ways to control surface acidity of inorganic fillers have been discussed by Solomon [34] and others [35]. Acidic groups fixed on the surface permit the filler particles to be treated with organic amines [36].

Other methods of filler treatment involve the coating of the filler particles with polymer. There are two approaches to this problem:

1. Pre-encapsulation of the filler to provide an immobilized polymer coating [37]; this may in principle be effected in different ways, e.g., by precipitation of the polymer from its solution or suspension, polymerization *in situ*, or mixing of the filler with a viscous polymer in mixing machines. (For a review of the patent literature in this field see [38].)

2. Monomer polymerization on the filler surface preactivated in one way or another; in this case either chemical bonds between the filler and the polymeric coating are formed, or a strong physical interaction results.

The first of these approaches has been effective only in a few cases; it will be considered in Section 3.4. Polymerization on active sites fixed on the solid surface opens, in our opinion, much better opportunities. This approach will be discussed in Sections 3.2 and 3.2.

3.2 Coating of the Filler Surface with Polymers by Radical Polymerization Techniques

If you ask an organic chemist about the best way to solve the problem of the interfacial layer between the polymer matrix and filler particles, he certainly suggests a chemical grafting of macromolecules to the solid surface.

The range of techniques available in solving this problem is very broad. Unfortunately, in many cases the extra cost of filler treatment overweighs the useful result. One example of economic failure is the story of "ceraplast" [39, 40], a polymer modified clay, which had been proposed for reinforcing polyethylene. The gain in quality was unquestionable, but the cost of ceraplast was more than four times that of unmodified clay. The process of the ceraplast synthesis included three major chemical steps. However, in recent years more simple processes have appeared, the economic attractiveness of which implants optimistic expectations.

It is commonly assumed that a treatment with a coupling agent should precede the coating of a filler surface with a polymeric film. Only in this way such adverse effects as filler agglomeration, side reactions on the particle surface, etc. may be avoided successfully. Such combined approach is generally taken, for instance, in the case of short and continuous reinforcement fibers.

Graft polymerization by the free radical mechanism will be discussed in more detail. The reason is that radical polymerization and copolymerization are, in many cases, independent upon the presence of moisture on the surface. Owing to this fact it is often unneccessary to subject the filler to a heat treatment, which is an energy-consuming and technically difficult operation. Several papers reviewing the state of the art in this field have appeared recently [41,42].

3.2.1 Physico-Chemical Initiation

Mechano-chemical procedures were among the first methods used for chemical grafting of macromolecules onto a solid surface. In one of the first works [43] a ball mill was used for grinding silica in a nitrogen atmosphere, in the presence of styrene, α-methylstyrene, butadiene, or chloroprene. But the amount of nonextractable polymer chemically bonded to the surface did not exceed 2% of the filler mass.

The application of ultrasound to metals of Groups III and VIII of the Periodic Table, in the presence of liquid monomers, resulted in surface polymerization [44]. No homopolymerization was observed.

Kargin and Tseitlin [45-47] have shown that ion-radicals are the active sites formed in the process of mineral grinding. The free radical mechanism of the physico-chemical initiation of polymerization has been established by Taubman et al. [48,49]. Although the degree of grafting depends upon the nature of the solid surface, the polymerization rate is practically independent of this factor. This conclusion was deduced from analysis of various dispersed materials including ionic crystals, metal oxides, carbonates, sulfates, amorphous glass, metals, graphite, polymers [50].

Negatively charged crystal centers [51,53] and, for organic fillers, charged surface defects may also be the active sites [52].

3.2.2 Radiation Grafting

Irradiation of mineral fillers initiates the polymerization of such monomers as styrene and methyl methacrylate [54,55]. The gamma-irradiated fillers have reactive sites which remain active through a period of 20 days.

The number of works devoted to this technique has been very large in recent years, but styrene and methyl methacrylate still remain the most favored monomers. Apart from the grafted polymer, homopolymer is always present in large amounts [56-58]. Fukano and Kageyama [59,60] carried out a systematic investigation of the gamma-initiated polymerization on solid surfaces. The monomer was adsorbed on an aerosil or silica gel surface from the gas phase, so that the monomer layer was very thin.

The filler was either irradiated alone, or irradiated in the presence of the monomer; the yield of grafted polymer was higher in the latter case. The grafted polymer was found to have a higher molecular weight than the homopolymer. In the presence of filler, the rate of polymerization increased which, according to these authors, was due to the excitation of the monomer molecules by way of energy transfer from the filler surface. One principal conclusion of this work was that radical and ionic polymerization processes are concurrent. This conclusion was drawn from the bimodal form of the molecular weight distribution.

Abkin et al. [61,62] assumed that surface silanol groups were responsible for the polymerization. They also showed [63] that the polymerization rate of vinyl acetate adsorbed on aerosil was different from that in the monomer bulk. An increase of the molecular weight of the grafted polymer as compared to the bulk polymer was attributed to the sharp decrease (five orders of magnitude) of the bimolecular termination constant.

Plasma may also be responsible for surface activation of mineral fillers [64]. Photoinitiated graft polymerization may be effected following treatment of the filler with peroxides [65], chlorsilanes [66], or quinones [67]. Apparently, the radiation causes decay of these compounds and the formation of grafted radicals which then initiate polymerization. The authors of the above papers used silica-based fillers and the radicals formed on the surface had the structure

$$\equiv Si^{\cdot} \quad \text{or} \quad \equiv Si—O^{\cdot}$$

The formation of homopolymer seems to be practically unavoidable in radiation initiation. In most experiments the polymerization took place predominantly in the sorbed layers.

Little has been reported so far concerning the practical use of radiation-initiated grafting of polymers to filler surfaces.

3.2.3 Grafting to Cellulosic Materials

Grafting to cellulose may be regarded as a special case of polymerization filling. Historically this method had been developed for the purpose of modifying cellulose in order to impart improved properties to it. Many reviews have been written since then [68,69]. Within the scope of this monograph we are interested primarily in cellulosic materials as fillers for plastics. Wood flour, powdered cellulose, or milled nutshells are example of commercial fillers of this type. A basic limitation to their use is their low heat resistance. A historic milestone was the creation of modified wood materials impregnated with phenol-formaldehyde resins. After World War II, as the radiation technology was progressing, radiation began to be used for initiating vinyl monomer polymerization. The pioneering works were carried out in USSR by Karpov et al. [70], who impregnated wood with methyl methacrylate and its derivatives or with styrene, and then exposed it to gamma irradiation. Almost simultaneously parallel work was undertaken in USA [71]. Electron beams and traditional chemical initiation techniques [72], as well as photoinitiation [74-76] have also been used for the purpose. However, radiation was found to be the most effective means of grafting polymers to the filler material surface [73].

The best way to modify cellulosic materials is by minor homogeneous swelling in the impregnating monomer. Polymer grafting to cellulose fibers has been widely applied in the manufacture of cotton articles. The methods developed in this industry may be used directly for the modification of cellulose dispersed in plastic matrices.

3.2.4 Fixation of Active Sites on Filler Surfaces

There are several principal techniques for the fixation of macromolecules on the filler surface involving chemical initiators [41]. They include:

a) grafting (adsorption) of initiators containing functional groups;
b) grafting of multifunctional copolymerizable compounds;
c) use of functional groups available on the surface for copolymerization with the corresponding monomers;
d) grafting of multifunctional groups followed by their conversion to initiators.

The graft polymerization of acrylonitrile, acrylic acid, methacrylic acid, methyl methacrylate, vinyl acetate on Al_2O_3, TiO_2, $CaCO_3$, ZnO, Al has been investigated [77]. The initiators (benzoyl peroxide or ammonium persulfate) were adsorbed on the filler surface, and the monomers were deposited from the liquid phase. The effects of the types of monomer, initiator and substrate are illustrated in Table 3.2.

The low efficiency of methyl methacrylate grafting under the given conditions, as well as the low activity of benzoyl peroxide are remarkable. It should be noted, however, that the latter is soluble in the monomers and can, therefore, be transported

Table 3.2. Effect of the Type of Monomer, Mineral Compound and Initiator on the Graft Polymer Yield α [70]

No	Filler	Filler surface, m^2/g	Initiator	Polymer	Polymer yield, %		α, %
					m_1	m_2	
1	Al_2O_3	93.80	APS	PAN	20.0	13.0	69.0
2	Al_2O_3	93.80	BP	PAN	0	0	0
3	Al_2O_3	93.80	APS	PAA	55.4	53.6	97.0
4	Al_2O_3	93.80	BP	PAA	10.7	8.0	73.0
5	Al_2O_3	93.80	APS	PMAA	30.0	26.4	87.6
6	Al_2O_3	119.0	APS	PAN	26.3	17.9	72.2
7	Al_2O_3	145.0	APS	PAN	43.4	38.4	88.5
8	$CaCO_3$	0.12	APS	PMAA	15.6	15.1	97.00
9	$CaCO_3$	0.12	APS	PAA	33.6	33.5	99.0
10	TiO_2	3.50	APS	PAN	22.8	3.5	15.4
11	TiO_2	3.50	APS	PMMA	12.4	1.0	8.8
12	ZnO	2.40	APS	PMMA	9.6	0.8	8.1
13	Talc	0.12	APS	PAA	8.3	3.7	44.2
14	Talc	0.12	BP	PMMA	27.0	2.7	10.0
15	Talc	0.12	BP	PVA	12.8	2.9	22.3

BP: benzoyl peroxide, PAN: polyacrylonitrile, PAA: poly(acrylic acid), PMAA: poly(methacrylic acid), PMMA: poly(methyl methacrylate), PVA: poly(vinyl acetate); m_1 and m_2 are the quantities of the polymer formed and extracted, in % of the substrate weight; α is the fraction of grafted polymer in the total polymer yield.

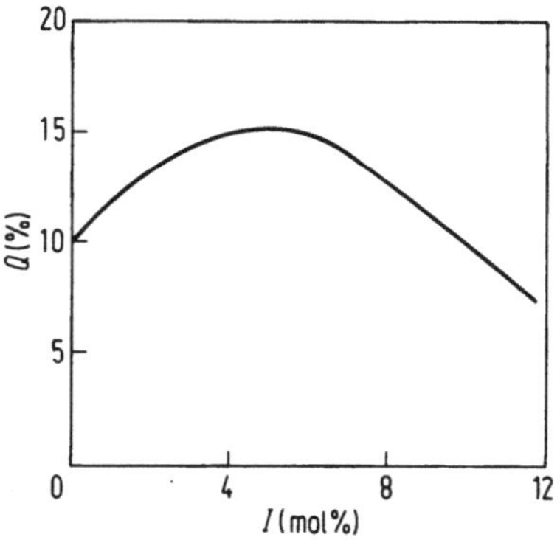

Fig. III.3. Dependence of the polymer yield on the initiator concentration for the system: MMA, talc and BP [70].

to the bulk, rather than remain on the filler surfaces. PMAA is also soluble in its monomer.

The data given in Fig. III.3 indicate an unusual relationship between polymer yield Q and initiator concentration I for the system MMA, BP, and talc. The authors [70] attribute this to chain transfer to the initiator, but do not substantiate this hypothesis by molecular weight determinations.

Alumosilicates and silica gel were used as components of redox initiation systems in emulsion polymerization [78, 79]. To this end, Fe^{2+} and Cr^{3+} ions were deposited onto the surface layer of the filler by ion-exchange reactions. In the presence of peroxides, redox systems were formed which initiated the polymerization of vinyl monomers. Grafted polymer amounted to 60–80% of the total polymer yield. Owing to the ion-exchange activity of most mineral fillers, the above redox initiation method appears to be quite versatile.

Initiators containing functional groups such as carboxyls may be fixed onto the filler surface by their reaction with OH-groups. In Japanese patents [80, 81] succinic acid peroxide, tert-butylperoxymaleic acid, their salts, and a number of other typical peroxides were used in combination with a carboxy-containing reducing agent. The fillers were added in various forms, as fibers, powders, films or sheets, and a large number of different filler materials was investigated, inorganic and organic solid materials, metal oxides and dispersed metals, salts, hydroxides, pigments. As in other cases, the polymers (PS, PMMA) formed consisted of a mixture of grafted and free polymer, the former making up about 50% of the total.

Mineral fillers with acidic properties can be effective for graft polymerization, provided one of the radical generating components contains a free or substituted amino-group [81]. Such an agent may be a single initiator (e.g., 2,2-azobis-(2-aminopropane)-hydrochloride), a redox system consisting of peroxide and amine, or a monomer (aminosubstituted acrylates, methacrylates, etc.). Although in these cases the grafted polymer is produced only by chain transfer, the authors claim grafted product yields of 80–90%.

Glass fibers have been boiled in water to obtain silanol groups on the surface by the reaction [82, 83]:

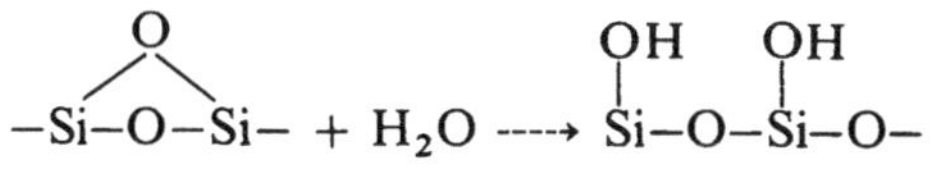

Their concentration was determined by chemical analysis [84]. Treatment with allylglycidyl ether permitted reactive C=C bonds to be fixed on the fiber surface:

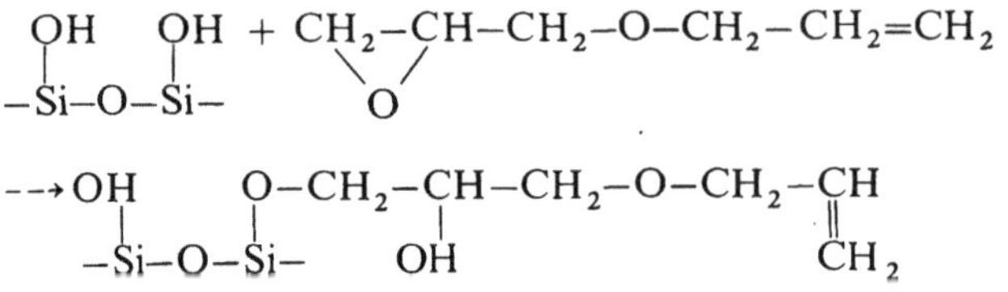

Graft polymerization of vinyl monomers, including methyl methacrylate, vinyl chloride [85], styrene, etc. may then be carried out using peroxides as initiators [86].

Implementation of a commercial process for grafting polystyrene to glass fibers has been reported by MITSUBISHI MONSANTO, Japan [87], however no details were disclosed. Modified fiberglass concentrates have also been used as reinforcements for ABS-plastic. A considerable gain in impact strength and tensile strength has been reported [87].

3.3 Ziegler-Natta Polymerization on Filler Surface

Radical graft-polymerization processes are not suitable for polyolefins such as LDPE, HDPE, polypropylene, ethylene-propylene copolymers, poly-α-butene, etc. The development of catalytic polymerization processes by the ionic-coordinative mechanism has led to a new generation of catalysts applied on organic or inorganic fillers. The use of Ziegler-Natta catalysts for the production of polymerization-filled polyethylene was suggested simultaneously and independently in the USSR [88-90] and USA [91-93]. These works gave impetus to a new trend in this field, and some of the results surpassed expectations.

3.3.1 Polymerization Processes

Table 3.3 summarizes the possible polymerizatic filling methods.

Today there exist processes to synthesize polyolefins on a very wide range of fillers, including minerals, metals and their oxides, graphite, carbon black, glass and organic fibers. In all cases the first step is filler conditioning. Depending on the nature of the filler, the conditioning has to be carried out differently, but in any case the aim is

Table 3.3. Conditions for grafting polyolefins onto fillers by Ziegler-Natta type catalysis.

Process conditions	Catalyst fixation method	Phase ratio
Liquid phase	Treatment with active components	Synthesis of filler "concentrates" coated with a polymer film
Gas phase	Use of active components on filler surface	Synthesis of filled + unfilled polymer mixtures; synthesis of graft polymer with a given filler concentration.

Table 3.4. Catalytic systems used for grafting polyolefins onto fillers.

$TiCl_4 + AlR_3$	$Ti(OR)_4 + AlR_nCl_{3-n}$	$TiO_2 \cdot ZnO \cdot AsO$
$TiCl_4 + AlR_nCl_{3-n}$		
$VCl_4 + AlR_3$	$VOCl_3 + AlR_nCl_{3-n}$	
$VCl_4 + AlR_nCl_{3-n}$		
$Cr(O{-}\overset{\overset{O}{\|}}{C}{-}R)_3$	$CrRCl_4$	$Cr(C_2H_5)_2$
ZrR_4		

to remove all polymerization inhibiting compounds from the surface [94]: H_2O, CO_2, SO_2, O_2, etc. Too many hydroxyl groups on the surface are also undesirable since they too inhibit the polymerization.

Filler drying at high temperature, in vacuo or in a current of an inert gas, is a traditional procedure in powder technology. For some fillers (mainly minerals such as perlite, kaolin, glass beads) the conditioning may include open firing. For fillers containing physically sorbed water, or crystalline hydrate water, it is known that heat treatment at different temperatures causes desorption of only a certain "layer" of moisture. Complete removal of water from aluminum hydroxide, kaolin, perlite, and other water-containing minerals is economically unfeasible. But, to carry out surface polymerization successfully, it is necessary only to control the content of OH-groups on the surface.

Dehydroxylation of the filler material surface may, alternatively, be effected by chemical methods, for example with aluminum alkyls. However, it was shown, in particular for the case of clay, that apart from the reaction with surface OH-groups, also some alkylation of siloxane, alumoxane and Si—O—Al groups took place.

Some types of catalytic systems that have been used for polymerization filling of polyolefins are listed in Table 3.4.

Since all these complex-based catalysts show acidic properties, their reactions with the essentially basic fillers would result in neutralization and deactivation of the active sites. But the fillers can be pre-treated with nonvolatile mineral acids, and then they will be suitable for use as polymerization carriers. One of the methods proposed consists in contacting minerals with an aqueous suspension of $Al(NO_3)_3$:

$$(CaCO_3)_n + 2\,Al(NO_3)_3 \xrightarrow{H_2O} Al_2O_3(CaCO_3)_{n-3} + 3\,Ca(NO_3)_2 + 3\,CO_2\uparrow$$

Such agents as SiO_2, Al_2O_3, H_3PO_4, and others may also be used for the purpose.

The first experiments using $TiCl_4$ + $Al(alkyl)_3$ as the catalyst for grafting of polyethylene onto cellulose fibers and carbon black were carried out in 1964–1965 [97,98], but the materials obtained were not sufficiently homogeneous. By now several methods of fixation of a transition metal compound on a filler surface have been developed:

Physical sorption

Sublimation, impregnation or deposition of transition metal compounds lead to the formation of a solid phase on the filler surface or in pores [94]. The process may involve either oxidation or reduction of the transition metal. Application of such a solid catalyst phase, which strongly adheres to the surface, is essential for obtaining an isotactic polymer in the polymerization of propylene and higher olefins.

Chemical reaction

Hydroxyl groups on the surface of the filler can react with a transition metal compound and an organometallic compound according to the reaction [99]:

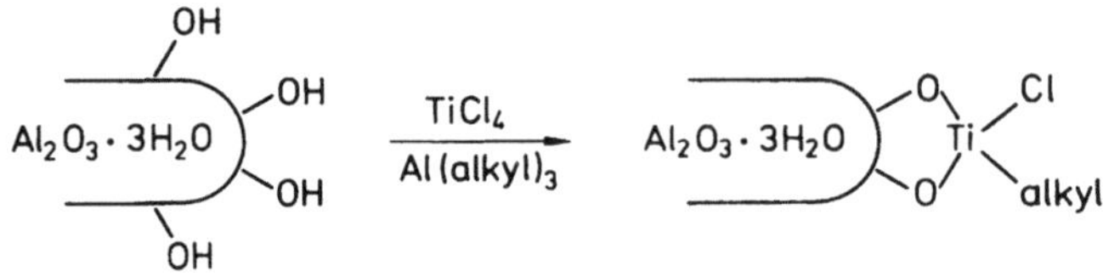

Single-component catalytic systems

Recently a series of deposited ethylene polymerization catalysts have been prepared by applying organic and hydride compounds of transition metals onto oxides [100, 101]. In the free form these compounds show only weak or no activity. But after reaction with the surface hydroxyl groups they become highly active, not requiring any further treatment with aluminum alkyls. The general formula of these surface products is

$$(Al(Si){-}O)_x{-}MR$$

where M is the transition metal, R is $-C_2H_5$, $-CH_2C_6H_5$, $-BH_4$ etc.

Standard alumina or silica carriers for ethylene polymerization catalysts have a well developed surface (200–300 m^2/g). Common oxide fillers, on the other hand, have a specific surface area of only $\simeq 10$ m^2/g. This difference has an influence on the kinetics of the polymerization. Logically, the polymer output is more for the first type of carriers, but unexpectedly, the molecular weight of polyethelene is higher for the second type.

Tetrabenzyltitanium was also used as a catalyst; active systems were obtained on asbestos, 2 CaO SiO_2, and on ground perlite [102]. In order to increasing the content of reactive OH-groups on the surface, the fillers were treated with $AlEt_3$ and then subjected to hydrolysis.

The use of the catalyst $Zr(PhCH_2)_4$ was disclosed [103]; the addition of some $Al(alkyl)_3$ permitted the Zr concentration to be reduced by a factor of 5–10, for carbonate fillers. Chromocene on a SiO_2 carrier was also used as a catalyst for ethylene graft polymerization [170]. But no specific data were published about the polymer properties and kinetics of the process.

Surface activation

The use of calcined clay (montmorillonite) as a catalytically active filler for the grafting of polyethylene has been patented [104]. Activation takes place during calcination at above 400 °C. The active component is assumed to be TiO_2 which is catalytically active in the presence of Al. The surface may, therefore, be pre-treated with TiO_2 and then subjected to hydrolysis and activation [105].

Kinetically, the process of the olefin polymerization on fillers is similar to that on heterogeneous catalysts. The observed decrease of polymerization rate with time is usually attributed to desactivation of the active sites and deterioration of the mass transfer conditions. However, it has been shown that if $Mg(OEt)_2$ is used for the activation of fixed $TiCl_4$, the rate curves become similar to those of the homogeneous ethylene polymerization (with an initial rate increase) [106]. The nature and conditioning method of the filler, as well as the type and method of application of the catalyst, affect the initial polymerization rate of ethylene and the shape of the kinetic curves.

The liquid-phase polymerization on finely dispersed fillers is experimentally difficult due to filler aggregation in hydrocarbons. To improve the homogeneity of the system, it was proposed to pretreat the filler with $Al(alkyl)_3$ [92]. The viscosity of a concentrated suspension was thereby reduced by about a factor of 10^3, and was almost the same as that of pure hydrocarbon.

The fluidized bed technique was found to be optimum for gas-phase polymerization [89]. The main difficulties in this process are associated with the filler particle size and the possibility of particle agglomeration at the early stages of the polymerization. Classical methods are applicable to particles whose average size is within 50 to 100 μm, but it is advantageous to use a fraction of particles of a smaller size in order to increase the stability and density of the fluidized bed. But this fraction should not be more than 10–15%. Ultrafine particles (1 to 10 μm), however, will destabilize the process and increase particle entrainment.

As an example we may refer to the UNIPOL process (Union Carbide Corp., USA), where the mean polymer particle size in the gas-phase polyethylene polymerization process is 100 to 300 μm. The catalyst is applied on silica gel particles (filler!) of 30 μm mean size, added to the reaction vessel in quantities not exceeding 0.05% of the polyethylene particles mass. This process is stable in large volume reactors. A corresponding fluidized bed process for the graft polymerization on fillers in the industrial scale was not developed, but calculations were made [113].

The kinetic parameters of the propylene polymerization with the catalytic system $TiCl_4 + (C_2H_5)_2AlCl$ deposited on perlite or graphite particles have been investigated [107]. The catalyst preparation conditions, the order in which the components were added, and their concentrations had a considerable effect on the initial activity of the catalyst and its time history.

More complex, crosslinked polymer structures may be obtained on filler surfaces by copolymerizing propanediene with olefins [108].

3.3.2 Molecular Characteristics of Composites Prepared by Graft Polymerization on Fillers

Graft polymerization of ethylene yields generally two types of macromolecules, viz. those chemically grafted to the filler surface, and homopolymer which is crystallizable in bulk. The latter is produced by a chain transfer mechanism.

Whatever the filler type and the particular catalytic system used, one factor which is common to all processes of ethylene polymerization on the filler surface is the formation of polymer having an ultra-high molecular weight (UHMWPE), somewhere in the range of $(0.6–7.0) \times 10^6$. The MWD of the graft polymer depends largely on the homogeneity of the polymerization system. In some cases, $\overline{M}_w/\overline{M}_n = 5.7$ to 7.4 [99], in other cases even between 10 and 20. The UHMWPE has a melting point of 137 to 138 °C, but after recrystallization it is 132–134 °C, which indicates a high degree of molecular orientation during the synthesis. It was shown that the degree of orientation corresponds to a stretching factor of 4–5; the initial crystallinity is 78–85% [94].

A closer investigation of the microstructure of polymerization-filled UHMWPE by the dilatometry technique [109] has shown that the crystallization conditions depend strongly on the filler particle type and size, and on its concentration by volume. The relationships observed were considerably different from those known for composites obtained by compounding.

Attempts to control the molecular weight of polyethylene during the polymerization on the filler surface by adding hydrogen, varying the catalytic system, and copolymerization with α-butene resulted in a sharp increase of product inhomogeneity and a deterioration of the mechanical properties. To date, the problem of reducing $\overline{M}_w$ cannot be considered solved.

Propylene graft polymerization was not accompanied by an increase of the molecular weight of the polymer [94]. The atactic phase content increased somewhat, up to a level of 9–14%. Catalysts obtained by low temperature reduction of $TiCl_4$ on the filler surface were the most stereospecific ones.

For vanadium catalysts, IR spectra showed the existence of chains resulting from isomerization polymerization ("head-to-head" addition). This causes considerable changes in the polypropylene properties, including an increase of the impact strength and a decrease of the brittle temperature for the atactic phase content (8 to 12%).

3.3.3 Rheological and Mechanical Properties of Composites Prepared by Graft Polymerization on Fillers

One of the main ideas behind polymerization filling was that this method may assure a high degree of homogeneity of the composite product. But speculations concerning the advantages of mixing filler particles with a low-viscosity monomer rather than with a high-viscosity polymer were not as founded as might be expected. The probability of formation of filler agglomerates or particles with nonuniform polymeric coating is rather high, both in gas-phase and in liquid-phase polymerization. Special analysis techniques and tests had to be developed for determining the structure parameters of these composites.

The somewhat unexpected results obtained in ethylene polymerization on dispersed fillers (see preceding section) apparently prompted most investigators in this field to channel their efforts in a different direction. The UHMWPE is a unique material. Melted it does not become a visco-elastic fluid, hence it cannot be processed by extrusion and injection molding. Superheated and subjected to high shear deformations it undergoes intensive degradation yielding material with a very broad MWD. The mechanical properties of degraded UHMWPE are quite inferior to those of HDPE with the same molecular weight. Thus, the idea of controlling the MW of polymerization filled polyethylene in the course of processing proved useless.

However, the pressing, sintering, compression and injection molding, and heat molding techniques applicable to pure UHMWPE are suitable for the filled product as well. Of course, UHMWPE cannot compete with commercial grades of HDPE in its traditional range of applications, but it possesses some unique properties which may warrant an ever increasing demand for it. They include high wear resistance, excellent chemical resistance and impact strength [110].

The UHMWPE may be filled with dispersed fillers by compounding techniques, but this is not very effective due to the very high viscosity of the molten polymer. Filling during polymerization is much more advantageous for the composite properties. UHMWPE synthesized by graft polymerization on a mineral filler, (filler content 50–75%) has an impact strength higher than that of unfilled polycarbonate or ABS-plastic in the temperature range from —40 to +30 °C [111]. Howard et al. [111] referred to the product as a "homogeneous composite". In the Soviet Union materials of this type are known with the trade name of Norplast [112].

American scientists focused their effort on the development of composites containing a maximum amount of filler (correspondingly having the maximum value of Young's modulus, as well as a high impact strength and ductility). A side line was to obtain a non-combustible material with a low smoke generation factor [110]. The problem was solved by a meticulous control of the particle dispersion in the liquid phase and the absence of free catalyst in the reaction system.

Such graft composites on the basis of $CaCO_3$ or $Al_2O_3 \cdot 3\,H_2O$ contained 56 to 72.5% of filler. For blended HDPE or UHMWPE composites, even with a coupling agent, such a high filler concentration would result in easy brittle fracture at a relative elongation of only 1.0%, the tensile strength would drop by 30–50%, and the Izod impact strength would decrease by about a factor of 10, as compared to unfilled polyethylene. The graft-polymerized composites, on the other hand, have an unusually high impact strength and a relative elongation at fracture of 30 to 100%, which makes them in fact ductile materials.

While for unfilled PE the Izod impact resistance increases with the molecular weight

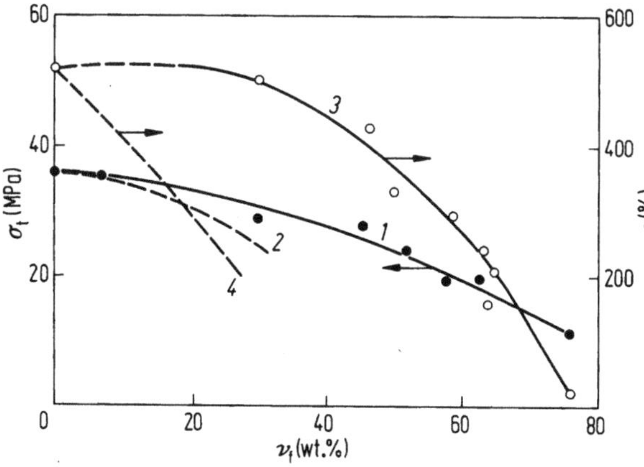

Fig. III.4. Dependence of σ_t and ε of composites with calcite (particle size 5 μm) on the weight fraction of the filler. *1*, *3*: graft-polymerized HMWPE composite; *2*, *4*: blended UHMWPE composite [182].

only up to 10^6, for graft-polymerized composites this parameter increased in the MW range 2×10^6 to 7×10^6, although the reasons are not quite clear yet (no drop of crystallinity was observed). Soviet data [94] corroborate these findings. Graft-polymerized composite specimens (with tufa, kaolin, calcite, Al_2O_3, milled perlite, etc. as fillers) filled to 50–60 wt% featured a relative elongation of about 100% at fracture. The dependence of tensile strength and relative elongation on v_f for these types of the composites in shown in Fig. III.4 [182].

Young's modulus of these materials is in good agreement with that calculated by the Einstein-Guth-Gold (EGG) equation [10]; for an $Al_2O_3 \cdot 3\ H_2O$ filler it is:

$$E_c = E_m(1 + 2.5v_f + 14.1v_f^2)$$

For specimens with clay filler E_c was somewhat overestimated by this equation (due to a filler shape effect).

In the stress-strain diagrams of low-filled pressed samples, there is generally a flow point and an orientation hardening segment. In the case of high-filled samples (>50 wt.%) both are not so apparent. In tension, the cross-sectional area decreases and microvoids are formed.

Interestingly, the use of kaolin filler particles of a smaller size (2 μm instead of 9 μm) increases the Izod impact strength, rupture strength, and relative elongation, whereas Young's modulus and the heat distortion decrease somewhat [99]. Similar results were obtained for tufa and kaolin fillers with a particle size range between 2 and 50 μm [113].

The authors of [113] investigated also the properties of perlite-vulcanic glass fillers and graft-polymerized composites thereof. Due to crystal water, calcination turns the perlite into a foam with a bulk mass of 40 to 60 kg/m³. Because of its high brittleness,

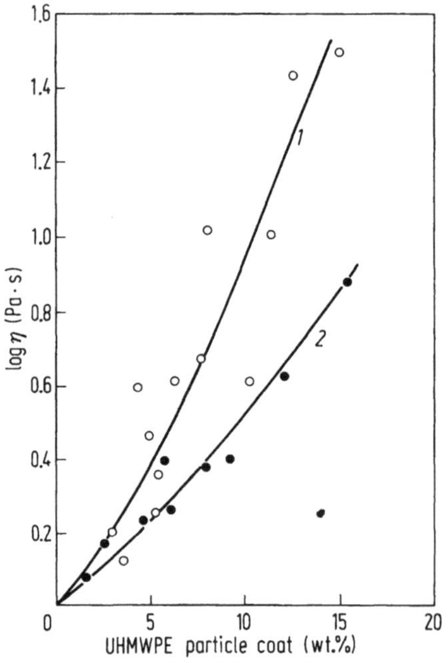

Fig. III.5. Dependence of the effective viscosity of the melt on the content of UHMWPE particle coats (wt.% of filler) in composites with 40 wt.% of perlite [115]. *1*: LDPE matrix; *2*: HDPE matrix;

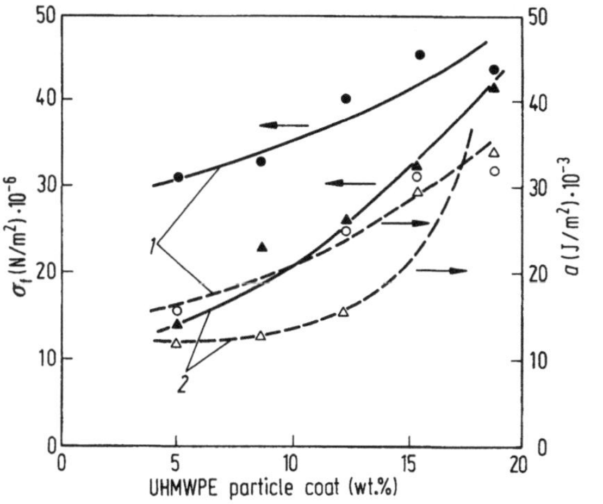

Fig. III.6. Dependence of mechanical properties of composites on the content of HMWPE particle coats (wt.% of filler). *1*: graft-polymerized composite; *2*: blended composite.

applications of this material are limited, except where its excellent heat insulation properties are important. Ethylene polymerization on its surface leads to a low bulk mass composite which is in fact a gas-mineral-filled UHMWPE. With a PE content of only 8–12 wt%, this material may be pressed into blocks with a bulk mass of 120–150 kg/m^3 and used as a heat and sound insulating material featuring a compressive strength of 0.4–0.8 MPa.

A coordinative polymerization process has been used for producing polyolefin films on glass fibers [118]. However, no comprehensive information about the mechanical properties of such glass fiber plastics is available so far.

Composites consisting of kaolin, tufa, or calcite, and 5 to 20 wt% of grafted polymer can be used as filler concentrates for mechanical blending with HDPE, LDPE, or PP [114]. The rheological properties of composites containing two HDPE grades with a flow index of 4.0 g/10 min and one LDPE grade with a flow index of 7.0 g/10 min were investigated with the aid of a Reograph capillary viscometer [115–117]. The relationship between the effective viscosity of the melt and the fraction of ultra high-molecular weight graft polymerized skin in composites filled with 40 wt% of perlite is shown in Fig. III.5 [115]. A comparison of experimental and theoretical relationships led to the conclusion that there is no mutual solubility between the graft polymerized skin and the polymer matrix, but there is some segmental diffusion. The composites were processed by extrusion and injection molding. Some of the foamed perlite particles were broken in the process and the density of the finished product was dependent upon the molding pressure. At 40 wt% of perlite it ranged from 0.9 to 1.14 g/cm^3.

A detailed analysis of the rheological behavior of compositions consisting of a graft-polymerized composite (perlite/UHMWPE) blended with polyolefin matrices has been published [119]. Figure III.6 shows the dependence of the tensile strength σ and the impact strength a of the compositions on the UHMWPE content in form of filler particle coats. Comparison of these data with those for a simple mechanical mixture of the same components (perlite, UHMWPE, PE) shows that the graft-polymerized skin has a positive effect if present in an amount of 7 to 15 wt% of the filler mass. The same compositions were also used to test the effect of filler concentration on the composite strength. Injection molded and extruded test samples show a sharp peak of strength for a filler content of about 40 wt %. The elongation at rupture of these composites is twice as high as that of the blended mixtures.

Special emphasis was laid on testing the fatigue properties of the graft-polymerized composites. It is known that dispersion-filled blended composites on the basis of thermoplastics tend to fail easily in such tests. This characteristic is responsible for their restricted use, e.g. as a material for high-pressure conduction pipes (no more than 7 % of filler permittable). The UHMWPE-based graft-polymerized composites have been shown to have a higher fatigue strength than blended composites [120].

Most of the data on the physico-mechanical properties of graft-polymerized composites published so far refer to polyethylene, or rather, UHMWPE. On the other hand, no UHMW polymers are formed on the fillers in the cases of copolymers of ethylene with α-butene or propylene, nor for propylene homopolymer and block-copolymers with ethylene. These systems have not been studied thoroughly as yet, but it has been reported [94] that, by polymerizing propylene with the help of vanadium compounds, one can obtain a high impact material even with 40 wt % of a rigid mineral filler. The brittle temperature is −35 °C, which is typical of block-copolymers with ethylene, or for polypropylene compounded with elastomers. The quite unusual balance of properties of this material further includes a high rigidity and a tensile strength similar to that of the unfilled material. Such a set of properties cannot be obtained by componding.

3.3.4 Physico-Chemical and other Properties of Composites Prepared by graft Polymerization on Fillers

For PE-based polymerized composites investigators have noted a higher thermal stability and the presence of induction periods in the thermooxidation curves [121].

The electric conductivity of graft-polymerized composites on the basis of graphite or carbon black, with UHMWPE or PP matrices, has been investigated [122]. A comparison with blended composites shows that for the same filler concentration the electric conductivity of graft-polymerized samples may be 10^2 to 10^8 times higher! The linear electroconductivity versus sample size relationship testifies to a high degree of homogeneity which apparently cannot be attained by the compounding methods.

Low-temperature (4.2 to 300 K) electroconductivity tests have shown still another feature of graft-polymerized composite behavior. In this temperature range a graphite filled sample did not show any dependence of the electric conductivity on the temperature, for graphite contents from 8 to 61 vol. %. Multiple heating-cooling cycles in the same temperature range did not affect the mechanical strength and plasticity [123].

For many technical applications, polymeric composites featuring a combination of good mechanical properties and an electric conductivity in the range from 1.0 to 10.0 $Ohm^{-1}cm^{-1}$ would be highly desirable [124]. Undoubtedly polymerization filling opens up new possibilities in this direction.

A still unanswered question refers to the nature of the polyolefin-filler surface bond in polymerized composites. According to current conceptions of coordination catalysis, the metal-carbon bond is quite labile and is easily hydrolyzed under the attack of polar agents. There is no reason to assume that the nature of this bond is radically changed by the fixation of the catalytic complex on the solid filler surface. Indeed, the grafted UHMWPE and PP may be completely dissolved and removed from the filler using conventional procedures.

Under normal service conditions one cannot rule out the possibility of water diffusion to the particle surface which inevitably will cause hydrolysis of the bonds. However, a strong adhesion between the polymer coat and the solid surface may develop during the polymerization process, in analogy with the „scale" formed on reactor and mixer walls. The surface polymerization also serves to improve the interphase adhesion, in a way which cannot be achieved either by blending with molten polymer or by polymerizing in the presence of filler with free catalyst.

3.4 In situ Polymerization

In the previous two sections we discussed the polymerization on filler surfaces to which active sites had previously been attached in one way or another. The resultant polymer is, at least partially and temporarily, chemically bonded by covalent bonds to this surface.

This does not exhaust the possible range of methods for forming a polymeric coat on the filler surface. Alternatives include the use of certain types of fillers as *ionic* polymerization catalysts for such monomers as styrene or formaldehyde.

Of a certain interest is also the impregnation of the filler particles with monomers followed by independent *in situ* polymerization of the latter. In some cases a very effective interaction between the polymer and the filler can be achieved in this way. This procedure was adopted in work aimed at imitating the structure of dental enamel, a natural high-filled composite of hydroxyapatite crystals in a polypeptide matrix [127]. The solid phase content in the enamel is above 95 wt %. A model composite was obtained by polymerization of methyl methacrylate in which ceramic particles were dispersed. A lithium alumosilicate was first treated with silane to improve dispersion, whereby the filler concentration could be raised to 75–83 wt %. The polymerization was radiation initiated, and the filler phase homogeneity was assured by centrifuging the reaction vessel. The flexural strength maximum was recorded at the maximum filler content (82.5 wt %). Characteristically, when no centrifuging was used, micro-

Table 3.5. Mechanical properties of Asterite and some construction materials [129]

Material	Flexural strength, GN/m^2	Relative impact strength (falling ball test), cm
Ceramic	19	40
Asterite	12	150
Fiber glass reinforced plastic (polyester)	10	700
Wood	9	80
Acrylic polymer	3	200

voids formed in the systems filled above 80 wt %, causing a sharp drop of the strength properties.

ASTERITE (ICI), which is a poly(methyl methacrylate) high-filled with SiO_2 (>60% filler) is produced by in situ polymerization [128]. Some of the properties of this composite are compared in Table 3.5 with those of other materials.

Although no synthesis details have been disclosed, it is obvious that the *in situ* polymerized composite has a high impact strength in combination with a high rigidity.

3.5 Matrix Modification

There are three major methods of matrix modification, viz.: blending of polymers, copolymerization, and grafting of functional groups. We are not considering here low molecular weight plasticizers.

Blending of polymers. In most cases, a mixture of two polymers is a two-phase system. Phase distribution and structure depend heavily on the blending conditions and processing techniques. Impact resistant mixtures of thermoplastics and elastomers have been considered in detail in Ch. 1, where interfacial additives have also been discussed. We mentioned there paraffin wax and ethylene oxide oligomers as filler modifiers, but their effect may also be regarded as a matrix modification. The same holds true for filler encapsulation in polymer skins. The latter may be obtained by polymerization (cf. Sections 3.2; 3.3) or recrystallization of the corresponding polymer from the melt or solution.

Even if no special techniques are used (such as polymerization on the filler, sequential blending of the filler with the components), the phase distribution in a two-phase matrix will not be random because of different affinity of the components for the filler material.

Although technologists very often use polymeric additives for obtaining a composite with particular required properties, detailed structural analysis is carried out very rarely. One example is the widespread practice of adding thermoplastics to filled polyester-based composites. The benefit expected from this is a reduced shrinkage in molding and a certain increase of impact strength. Another example is the addition of LDPE to HDPE with calcium carbonate as a filler, for the purpose of reducing the composite brittleness [130]. HDPE has been widely used for improving the impact resistance characteristics of filled PP [131, 132]. In all cases the added polymer forms heterogeneous structures (domains), with sizes varying from fractions of a μm to 5 μm.

Commercially manufactured filled polymer mixtures include: polyamide/elastomer/glass fiber; PBT/PET/glass fiber; PBT/PET/mineral filler; PET/polyacrylate/glass fiber; poly(phenylene sulfone)/HDPE/mineral filler; poly(phenylene sulfide)/teflon/glass fibers [137].

Copolymerization of styrene with maleic anhydride ensures a better matrix adhesion to most fillers. Propylene copolymerization with small amounts of acrylic acid (0.3 to 6.0%) is very effective for glass fibers fillers. Such materials, as well as glass fiber reinforced copolymers of PP and maleic anhydride, are produced commercially [184].

Grafting of functional groups. In a general form, the effect of polar groups in the matrix polymer on the composite properties has been treated by Ford and others [134].

They concluded that the polarity of the matrix polymer (measured by the dielectric constant) is the main physical factor of the reinforcing efficiency of thermoplastic resins.

In Japanese patents [135, 136] it was proposed to graft glycidyl acrylate or methacrylate on polyolefins and polystyrene, to improve their adhesion to the filler (glass fibers). Tensile and impact strength of the composites were thereby improved, and the thermal deformation point was sharply increased.

3.6 Processing of Filled Polymers

The main approach to the shaping of articles from filled thermoplastics is the adaptation of available injection molding, extrusion, and thermoforming equipment for the purpose. The idea underlying this approach is to make the potential buyer believe that the transition from unfilled to filled plastics would be smooth.

Taking this strategy for granted we should not, however, disregard the specifics of processing filled materials. Their extensive use may, with time, bring about a considerable revision of the processing methods and equipment used today.

3.6.1 Injection Molding

Although considerable progress has been achieved in processing filled thermoplastics into sheets, pipes, profiles, and even films, injection molding remains still the major technology.

Since glass-filled thermoplastics may be singled out as a particularly promising group of materials, we will discuss briefly the main requirements to their processing:

1) The molding conditions must be such as to avoid fiber fracture;
2) rigid mineral and synthetic fillers usually increase the viscosity of the melt and, therefore, a higher injection pressure or a higher pressure in the injection cylinder is required;
3) special requirements are imposed on the molds since the flow pattern has a very strong effect on the mechanical properties of the finished products.

To prevent the fracture of fibers, or to limit it at least to the lowest degree possible, the recommended practice is to use moderate injection rates, low screw rotation speed in screw-type molding machines, and moderate shear rates [171,174].

As we have pointed out in section 3.3.3, the addition of filler tends to make the molten polymer behavior more non-Newtonian; however, the necessity to use moderate shear rates makes it impossible for the manufacturer to decrease the effective viscosity of the melt by using high shear rates. Another possibility would be to raise the processing temperature. But this is always risky due to the tendency of the polymers to undergo rapid degradation at high temperatures.

Thus, given the existing generation of injection molding equipment, the combination of the highest possible injection pressure and an accurately controlled temperature is the main technological requirement for a successful processing of glass-filled (and also for dispersion filled) thermoplastics.

Considering principally new technologies, we may point to the possibility of applying low-amplitude ultrasonic shear vibrations (about 2×10^4 Hz) to the shaping tool [8]. Using HF techniques, it is possible to affect the viscoelastic properties of the composites in such a manner that at a certain specific (critical) frequency one may expect a sudden drop of the effective viscosity [139].

It has been proposed to use stratified polymer flow for intensifying the molding of high-viscosity polymeric systems [140, 141]. This technique is used for the co-extrusion of two- or multilayer films, sheets and fibers. When high-filled systems are to be processed, a two-layer flow is created in the mold channel, with a polymer of lower viscosity flowing in the outer layer adjacent to the channel walls. There are different ways to create such flow patterns, for instance by providing a forced feed of the low molecular weight polymer component to the dead zones of the inlet section of the molding instrument. Another possibility is to mix the low molecular weight polymer with the composite; it may diffuse to the surface in the course of the process. As an example [144], if polymerization-filled polyethylene (filler: tufa), pre-mixed with a low-viscosity polyethylene was processed, a filler-free outer skin of 200–300 μm was observed on the finished article.

The effective viscosity of the melt can be reduced by applying a combined shear by different methods [145]. This approach was used for asbestos-filled materials processed by extrusion, but it may prove promising for injection molding of filled plastics as well.

3.6.2 Effect of the Orientation in Flowing melt on the Anisotropy of the Finished Article Properties

One of the greatest advantages of fiber- or lamellae-filled composites is their improved heat resistance. This parameter does not depend on fiber orientation, but strongly depends on the degree of crystallinity of the matrix (Fig. III.7) [180]. As regards amorphous polymers, the glass transition point is the upper temperature limit of their usability, and the filler (including a fibrous type) cannot improve this situation. In contrast, for crystalline polymers the effect of fillers is very tangible, and composites based on such polymers may be used at temperatures close to the melting point of the thermoplastic matrix, provided the loads are moderate.

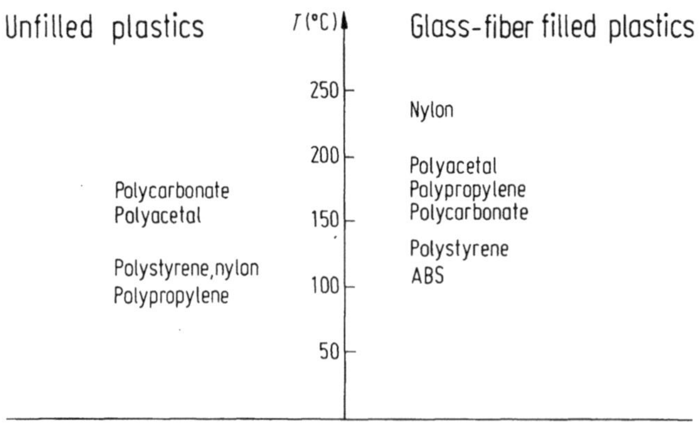

Fig. III.7. Effect of glass fiber filling on the heat resistance of thermoplastics [180].

The viscoelastic characteristics, viz. the tensile and flexural modulus of elasticity, are also little affected by fiber orientation, but they depend strongly on fiber dispersibility and size distribution.

One of the most difficult problems faced by a designer of plastic products is that of calculating the acceptable article strength. Designers usually rely on test results obtained by standard procedures. In test samples, glass fibers are predominantly oriented along the axis of bars or blades. It is quite obvious that in articles of complex shape, or in articles of simple shape to which stresses are applied in the direction transverse to the fiber orientation, the strength will be lower. It has been proposed to quote upper and lower strength limits for glass-filled thermoplastics, depending on the direction in which the load will be applied [146]. The authors studied the behavior of typical glass fiber filled polypropylene and PBT. Samples were cut in the direction of flow, and perpendicular to it. The creep modulus of glass fiber filled polypropylene (25% filler) measured on the two types of samples differed by a factor of 1.6–1.8, the tensile strength by a factor of 2.6. The strength of the specimen cut perpendicular to the flow was found to be the same as that of unfilled polypropylene. This coincidence may be accidental, but the difference between the upper and lower strength limits is indeed very high. In materials containing 15 vol. % of glass fibers, Young's modulus differed by a factor of 1.6–2.0 between the directions along and across the fibers. For glass fiber filled PBT the upper and lower strength limits measured in creep tests were also different by a factor of 2, and the discrepancy remained at elevated temperatures.

In a recent paper [147], the effect of the injection molding conditions on the properties of glass fiber filled polypropylene were analyzed in terms of the accompanying morphological changes in the matrix. The authors compounded glass fibers (28 wt %) with polypropylene on a laboratory screw extruder, without coupling agent. (The fiber size distribution in the extrudate and molded specimens was not studied.) Four injection molding process parameters were varied successively: injection pressure, injection rate, screw rotation speed and melt temperature. Mechanical tests were carried out be ASTM procedures. It must be noted that polypropylene is characterized by a very complex morphology, which is affected by inclusions in quite an unpredictable manner. In particular, the effect of injection pressure variations was quite unexpected. When the pressure was increased from 700 to 1300 kg/cm^2, both Young's modulus and the tensile strength decreased. It had been shown earlier [148] that, as the injection pressure increases, the degree of fiber orientation at the flow center (core) decreases, due to variation of the effective viscosity distribution across the flow. Detachment of the fibers from the matrix was also observed [147]. (This could have been due to the absence of coupling agents.) At low pressures surface nucleation and a subsequent transcrystalline morphology were observed. Such zones are known to improve the mechanical properties of fiber-reinforced polymer systems [179]. Despite the qualitative nature of these results, it would seem that the choice of the optimum injection pressure should be based on experimental runs at different pressures.

A three-fold increase of the injection rate (and consequently of the shear rate) improved the degree of orientation and increased σ_t. An increase of the injection temperature from 240 to 260 °C improved the fiber dispersion conditions. However, back in the early seventies it was established that injection molded specimens even of unfilled polymers must be considered as composite materials with a laminar structure [150]. The area fraction of skin tends to increase with the temperature. This skin does not have a spherolitic structure. Furthermore, the higher the melt temperature the higher the cooling rate in the mold. Due to the higher heat conductivity of polymers filled with organic fillers, their cooling rate in the mold is 20–30% higher, all other factors ignored. Thus, an increase of the injection temperature will give rise to higher relaxation stresses in the finished articles. Moreover, strength and Young's moduli are higher at lower temperatures.

The anisotropy of properties has also been investigated with short carbon fiber-reinforced polycarbonate specimens [151]. The problems involved are largely similar to those in the case of glass fiber reinforced thermoplastics, even though the nature of the filler and the matrix (amorphous polymer) are different.

Strength measurements at 90° to the flow showed a decrease by a factor of two. Young's modulus was measured by a procedure [152] involving three-point loading, at different angles relative to the injection flow direction. Special attention was paid to the effect of the weld-line on the mechanical properties of the specimens. The weld-line formation had little effect on the molded specimens if it was paralled to the axis of testing. But the average reduction in strength at an angle of 90° was about 50% (equal to strength of the matrix). The strength decreased as the injection pressure increased. At higher temperature the strength was also lower. Using the gel permeation chromatography technique, the authors showed that this is accompanied by degradation with reduction of the molecular weight of the matrix. This situation is not typical for glass fiber filled thermoplastics. For example, experiments with glass fiber filled polypropylene showed that the effect of an increase of the injection rate on the strength was adverse [153].

The ultimate goal of such investigations is, evidently, to obtain at least empirical equations, permitting to optimize the process conditions with the help of a computer.

3.6.3 Extrusion

The composite materials which are most often processed by extrusion are chalk-filled cable plastic (soft PVC), and chalk-filled rigid PVC for pipes of different types and uses. At least three circumstances are responsible for this selection:

1) availability of standard compounding equipment for PVC extrusion in existing production lines;
2) polarity of PVC, responsible for the strong "acid-base" interaction with the dispersed filler of basic nature (calcium carbonate);
3) obvious economic benefit from PVC filling (reduced cost per unit mass or volume of the composite, improved service properties or performance of the material, better machinability).

While the higher heat conductivity of polymers filled with organic fillers leads to a high heating-cooling resistance in injection molded articles, in the case of extrusion other factors are important. They include a better slip index, reduced risk of superheating at certain points in the melting zone, and decreased danger of "plate-out" in the extruder.

As regards the selection of the filler, practically all manufacturers arrived at similar results. For instance, quartz flour, silicic acid salts and asbestos cause too much wear in the working parts of the extruder; feldspar is too heavy, talc and kaolin render the compounded material too slippery and thereby make its processing irreproducible. Calcium carbonate in the form of finely dispersed chalk was found to be the most suitable filler. Most producers treat chalk with stearic acid, or its salt, to improve dispersibility and water resistance.

Natural chalk grades give, after milling and separation, irregular-shaped particles with a DES (diameter of equivalent sphere) of up to 2 μm. There has been a steady tendency for reducing the mean particle size of the filler, even at the expense of high extra cost. Larger particles have a strong adverse effect on the specific impact strength of the composite. Only some 10–15 years ago, chalk with a mean particle size of 20 μm was considered quite appropriate. But the bulk of chalk used for the purpose currently has a particle size between 2 and 5 μm, and 10 μm is already regarded as inacceptable. The chalk intended for compounding is precipitated chemically and has a mean particle size of 2 μm [154].

The filler content in the composite depends on its intended use. For cable sheath plastics, 20 to 25 % by weight are added; for drain, electric, ventilation and form pipes, 20 to 40 % by weight are generally used.

The filler content in material used for the manufacture of pressure, sewage and water pipes is, in many countries, regulated by codes at a level of 5–7 %. The reason is that PVC and polyethylene pipes filled to a higher concentration with a dispersed filler do not pass the standard tests for long-term resistance to internal water pressure. The use of various coupling agents has no effect on that behavior. However, three ways to overcome this obstacle may be outlined:

1) production of two-layer pipes in which the inner layer, which is in contact with the fluid, is made of an unfilled polymer;
2) crosslinking of the matrix polymer by physical or chemical means;
3) use of materials prepared by graft polymerization on fillers (see Section 3.3).

Any of the above methods involves higher production costs. Therefore, a careful market situation vs. cost analysis is needed before deciding upon any of them.

The addition of coupling agents and of special-purpose agents to extrudable materials has been extensively advertised as a means of improving the specific impact strength of polymer composites. Such agents include chlorinated polyethylene for PVC filled composites.

It has been shown, that calcium carbonate absorbs 40 % less plasticizer after a surface treatment with organic compounds [183]. A 25 % reduction of stabilizer sorption by calcium carbonate after treatment with stearates has also been reported [156].

Apart from PVC, dispersion-filled polyolefins are usable for extrusion, though on a limited scale. The combination of a mineral filler and foaming has been used by ARMOSIT, France, in their ARMOSEL tube extrusion process. While the density of rigid PVC is 1.4, that of the material filled to 25 wt % with $CaCO_3$ and foamed is 0.9 g/cm^3. Materials of this structure have been produced from LDPE (+4 % $Al(OH)_3$ + Freon) [155] and other thermoplastics [179].

Structural foam filled with fibers is a new product for extrusion, with a great potential market. The role of glass fiber reinforcement in foamed polyethylene has been demonstrated [179]. Surprisingly, glass fiber reinforced (15 %) foamed HDPE with 18 % voids has shown better properties than unfilled unfoamed polyethylene. Both materials have the same density. Applications for this type of filled-foamed material are being developed in many research centers. Reinforcement involving prefoaming

Table 3.6. Filled polymers (%) for floor coverings and decorative tiles [184]

Polymer Matrix	1979	1985 (forecast)
PVC	48	42
Polyester resins	33	38
Polypropylene	4	5
Phenolic resins	5	3
Epoxy resins	3	4
Polyamides	2	3
Other	5	5

and filling with two or more types of fillers is a very interesting direction for new developments.

Chalk filled PVC composites dominate by far over other filled polymer materials. Table 3.6 illustrates the predominant use of filled (30–80 wt%) PVC for the manufacture of floor coverings and decorative tiles.

3.6.4 Thermoforming

In the early 70's a commercial process was developed for the production of sheet materials from glass fiber reinforced thermoplastics, and for machining them into parts by stamping [138]. Although the production of sheets is generally a rather costly process, the stamping technique is simple, fast and cheap. It is especially due to rapid forming and cooling cycles that this method may prove competitive with the sheet molding compound (SMC) method.

Glass mats are used for this material. The polymer (as a binder) is applied as a suspension, solution, or powder. The system is processed by molding, calendering, and extrusion. The polymer matrix is usually polypropylene or nylon-6. The non-woven glass mats impart a high rigidity and impact strength to the material. The glass fibers, alone or in combination with a mineral filler, are usually added to the matrix polymer to improve its heat resistance. The advantage of this process over molding is that the length of the reinforcement fiber is unaffected.

Thermoforming is also used for processing dispersion-filled (with chalk or talk) polypropylene into large-sized articles. In this case the sheet material is obtained by extrusion. The sheet is first heated and then rapidly pressed between cold dies [181]. It might seem that any filled material can be processed in this way. However, the problem apparently lies in the difficulty of obtaining suitable sheets.

3.7 Physico-Chemical Properties of Polymer Composites

Since fillers are known to affect significantly the physico-chemical properties of matrix polymers, considerable variations in their thermal stability and susceptibility to thermooxidation may be expected. There have been only a few research papers and reviews published so far on this subject [157]. One is surprised to find strikingly different results in different published works. The easiest way to explain it is that the samples were prepared in different ways and contained uncontrollable admixtures.

The relation between the specimen preparation conditions and observed thermal effects has been studied specifically [158]. Using different methods of blending talc and dispersed iron with HDPE, the authors have shown that as the degree of interphase contact was increased the polymer melting and oxidation peaks were shifted to the low temperature region.

Fillers may be divided into those behaving inertly or actively in degradation processes. The former do not undergo any chemical or physical changes in the temperature range normally used for processing filled polymers, or occurring during their application, nor do they react with the polymer matrix. The latter may actively interact with the polymer matrix. The first group is usually believed to include such fillers as glass, talc, silica, and titanium dioxide. Highly dispersed metals, certain metal oxides, calcium carbonate, kaolin, and carbon black belong to the second group.

The apparent chemical inertness of polyolefins may give rise to an illusion about their stability in filled systems. However, the filler may either catalyze or inhibit the thermal or thermooxidative degradation of the polymer [159, 160].

The catalytic effect of dispersed metals on the thermooxidative degradation of HDPE has been ascertained for copper, nickel, titanium, iron, and lead [161]. In a later work it was suggested [162] that, according to IR spectroscopic data, fatty acid salts are formed on the metal surfaces. But it has been shown [163] that stearic acid salts have a stabilizing effect on the thermooxidative degradation of HDPE. So the mechanism of the catalytic effect of dispersed metals is unclear.

Glass fibers and spheres show no effect on the thermal and thermooxidative stability of polyolefins. It has been recommended to stabilize composites containing dispersed fillers such as talc, or asbestos fibers, with a combination of an antioxidant and a metal desactivator [164]. The degradation of filled vinyl polymers [165-167], and specifically of filled PVC [168, 169], has been widely investigated. Blending of metal oxides to PVC affects the yields and yield ratios of the different degradation products, although the qualitative composition of the products remains the same.

References

1. Van Oene, H., Ch. 7 in Polymer Blends (ed. by D. R. Paul, S. Newnan), vol. 1 New York: Academic Press 1978
2. Vinogradov, G., Malkin, A.: Polymer Rheology Moscow: Mir Publ.-Springer 1980
3. Thomas, D. G.: J. Colloid Sci. *20*, 267 (1965)
4. Menges, G. et al.: Kunstoffe *7*, 485 (1979)
5. White, J. L., Czaznecki, L., Tanaka, H.: Rubber Chem. Technol. *53*, 823 (1980).
6. Shenoy, A. V., Saini, D. R., Nadkarni, V. D.: Polym. Compos. *4*, 47 (1983)
7. Shenoy, A. V., Saini, D. R., Nadkarni, V. D.: Rheol. Acta, accepted for publication (1983)
8. Peshikovskii, S. L. et al.: Polym. Compos. *4*, 126 (1983)
9. Casale, A., Porter, R. S.: Polymer Stress Reactions, New York: Academic Press 1979
10. Goldberg, V. M. et al.: Dokl. AN SSSR *209*, 411 (1973)
11. Minsker, K. S. et al.: Plast. Massy No 9, 52 (1972)
12. Erdmenger, R.: DE-PS 862, 668 (1944)
13. Erdmenger, R.: DE-PS 813, 154 (1949)
14. Schneider, W., Zettler, H. D., Jeckel, G.: Verfahrenstechnik *12*, 477 (1978)
15. Schurr, V., Neumann, E.: Plastverarbeiter *29*, 351 (1978)

16. McNally, D., Freed, W. T.: SPE 32nd ANTEC (1974) p. 79.
17. Moskal, E. A.: Plast. Design Process *17*, 10 (1977)
18. Lu, S.-Z.: Proc. 35, SPI RP/Composites Conf. Section, 3-C, 1980
19. Frissel, P., Fillers, in: Encyclopedia of Polymer Sci. & Technology, v. 6, New York: John Wiley 1967, p. 740
20. Tager, A. A., et al.: Vysokomol. Soed. *A18*, 2201 (1976)
21. Tager, A. A., Jushkova, S.: Thes. Nat. Sci. Technol. Conf. Filled Polym. Mater., Moscow, NIITEKhIM, 1982, p. 20
22. Fowkes, F., McCarthy, D. C., Wolfe, J. A.: ACS Polym. Prepr. *24*, 228 (1983)
23. Motoyoshi, M.: Japan Plast. Age *13* (5), 33 (1975)
24. Plueddmann, E. P.: ACS Polym. Prepr. *24*, 196 (1983)
25. Plueddmann, E. P. (Ed.): Interfaces in Polymer Matrix Composites, New York: Academic Press 1974
26. Plueddmann, E. P.: Silane Coupling Agents, London: Plenum Press 1982
27. Ishida, H.: Polym. Prepr. *24*, 198 (1983)
28. Han, C. D. et al.: Polym. Eng. Sci. *21*, 196 (1981)
29. Nakatsuka, T. et al.: J. Polym. Sci. *24*, 1985 (1979)
30. Kokubo, M. et al.: Kubunshi Konbunshu *38*, 201 (1981)
31. Monte, S. J., Sugerman, G.: Proc. 35, SPI RP/Composites Conf., Sect. 1980
32. Mod. Plast. Intern. *11*, No 7, 29 (1981)
33. Mod. Plast. *56*, No 7, 42 (1979)
34. Solomon, D. H. et al.: J. Macromol. Sci. *A8* (3), 649 (1974)
35. Harvthorn, D., Solomon, D. H.: ibid. 653 (1974)
36. Solomon, D. H.: Brit. Pat. 1,228,538 (1969)
37. Hausslein, R. W., Fallick, G.: Appl. Polym. Symp. *11*, 119 (1969)
38. Atsuta, M., Turner, D. T.: Polym. Compos. *3*, 83 (1982)
39. Fallick, G. et al.: Mod. Plast. *45* (5), 143 (1968)
40. US Pat. No. 3,471, 439 (1969) to Bixler, D. and Fallick, G.
41. Ivanchev, S. S., Dmitrenko, A. B.: Uspekhi Khimii *51*, 1178 (1982)
42. Krocer, J., Shneider, M.: ibid. *43*, 349 (1974)
43. US Pat. 2,728,732 (1955)
44. Soma, D., Tabata, M., Kurosava, N.: J. P. Presentation No 53-143502 (1978)
45. Kargin, V. A., Plate, N. A.: J. Polym. Sci. *52*, 155 (1961)
46. Ceitlin, B. L.: Industrial Uses of Large Radiation Sources, Vienna: IAIE, 1963
47. Ceitlin, B. L. et al.: Radiation Chemistry of Polymers, Moscow: Nauka, 1973
48. Taubman, A. B., Janova, L. P., Blyskoch, G. S.: J. Polym. Sci. *9*, 27 (1971)
49. Janova, L. P. et al.: Kolloid. Zhurnal (USSR) *37*, 825 (1975)
50. Taubman, A. B.: Materials V National Symp. Mechanoemission and Mechanics (USSR), Tallin, v. 1, 1977, p. 79
51. Kuznetsow, V., Janova, L. P., Tolstaya, S.: Plaste u. Kautsch. No 3, 134 (1979)
52. Heinicke, G. et al.: Chem. *B19*, 118 (1979)
53. Eifler, R.-J. et al.: Ibid. *20*, 296 (1980)
54. Dinh-Ngoc, B., et al.: Angew. makromol. Chem. *B46*, 23 (1975)
55. Fukano, K., Kageyama, E.: J. Polym. Sci. Polym. Chem. Ed., *14*, 2193 (1976)
56. Blumenstein, A., Billmeyer, F. W.: J. Polym. Sci. *A4*, 465 (1966)
57. Steinberg, M. et al.: in: Multicomponent Polymer Systems (ed. by Platzer, N. A.) Adv. Chem. *99*, 547 (1971)
58. Abell, A. K., Crenshaw, M. A., Turner, D. T.: 179th A.C.S. Meeting, Houston, Texas, Organic Coatings Plast. Chem. Repr. *42*, 192 (1980)
59. Fukano, K., Kageyama, E.: J. Polym. Sci. Chem. Ed. *13*, 1325 (1975)
60. Fukano, K., Kageyama, E.: Tokyo soda Kenkyu Nokoky Sci., Rept. Tokyo Soda Manfg. Co. *22*, 3 (1978)
61. Mund, S. L., Bruk, M. A., Abkin, A. D.: Vysokomol. Soed. *A18*, 2631 (1976)
62. Bruk, M. A., Mund, S. L., Abkin, A. D.: ibid. *A19*, 889 (1977)
63. Bruk, M. A., Pavlov, S. A., Abkin, A. D.: Radiat. Phys. Chem. *17*, 113 (1981)
64. Berg, D. et al.: J. Chem. *18* (6), 219 (1978)
65. Negievich, L. A.: Ukr. Khim. Zhurnal *43*, 176 (1977)

66. Negievich, L. A.: ibid. *41*, 101 (1975)
67. Licov, N. I., Kachan, A. A.: Vysokomol. Soed. *B18*, 182 (1976)
68. Manson, J. A., Sperling, L. H.: Polymer Blends and Composites, London: Plenum Press 1976
69. Mishra, M. K.: J. Macromol. Sci.-Revs. Macromol. Chem. Phys. *C22* (3), 471 (1982—83)
70. Karpov, V. L., Kalinsky, V. M. et al.: Nucleonics *18* (3), 88 (1960)
71. Anon/ibid. *20* (3), 94 (1962)
72. Tailor, G. B., Dietz, G. R., Duff, R. M.: Nuclear Technol. *13*, 72 (1972)
73. Siau, J. E., Meyer, J. A.: For. Prod. J. *16* (8), 47 (1966)
74. Kubota, H., Ogiwara, Y.: J. Appl. Polym. Sci. *14*, 2879 (1970)
75. Nayak, P. L., Lenka, S., Mishra, M. K.: ibid. *26*, 2769 (1981)
76. Lenka, S., Nayak, P. L.: ibid. *27*, 3782 (1982)
77. Petrik, V. N., Kuznetsova, L. L., Lifshits, P. M.: Vysokomol. Soed, *B16*, 110 (1974)
78. Tutaeva, N. L., Komarov, V. S.: USSR Inventor's Certificate No. 444,780 (1972)
79. Tutaeva. N. L., Komarov, V. S.: USSR Inventor's Certificate No. 444,780 (1972)
80. J. Pat. Presentation No. 53–4028 (1978)
81. J. Pat. Presentation No. 53–4029 (1978)
82. Hashimoto, K. et al.: J. Appl. Polym. Sci. *27*, 4529 (1982)
83. Davidov, B., Kiselev, A., Zhuravlev, L.: Trans. Farad. Soc. *60*, 2254–1964)
84. Boehm, H. P., Schneider, M.: Z. Anorg. Alden Chem. *301*, 326 (1959)
85. Imoto, M., Minoura, J., Hayashi, T.: High Polym. Jpn. *15*, 26 (1958)
86. Maekawa, T. et al.: ibid. *21*, 321 -1964)
87. Mod. Plast. Intern. *8* (2), 22 (1978)
88. Enikolopian, N. S. et al.: USSR Inventor's Cert. 763,379 (1976); US Pat. 4,241,112 (1980)
89. Kostandov, L. A. et al.: USSR Inventor's Cert. No. 787,394 (1976); US Pat. 4,241,138 (1980)
90. Enikolopian, N. S., Volfson, S. A.: Plastmassy No 1, 39 (1978)
91. Lipscomb, R. D.: US Pat. 3,950,303 (1976)
92. Howard, E. G.: US Pal. 4,104,243 (1978)
93. Howard, E. G.: US Pat. 4,097,447 (1978)
94. Dyachkovsky, F. S., Novokshonova, L. A.: Uspekhi Khimii, *71*, N2, 200 (1984)
95. Chernaya, L. I., Matkovsky, P. E. et al.: Proc. IX Symp. on Synthesis and Properties Polymerized-Filled Polyolefins, Chernogolovka, Moscow District, p. 33 (1982)
96. Howard, E. G.: US Pat. 4,187,210 (1980)
97. Orsino, J. A., Hermam, D. F., Brancato, J.: US Pat. 3,121,698 (1964)
98. Herman, D. F. et al.: J. Polym. Sci. *11*, 75 (1965)
99. Howard, E. G. et al.: Ind, Eng. Chem. Prod. Res. Div. *20*, 421 (1981)
100. Ballard, D. G. H.: Adv. Cat. *23*, 163 (1973)
101. Ballard, D. G. H. et al.: polymer *15*, 169 (1974)
102. Ermakov, Yu., Zakharov, V. A.: Adv. Cat. *24*, 173 (1975)
103. Howard, E. G.: US Pat. 4,187,210 (1980)
104. Aldeman, R. L., Howard, E. G.: US Pat. 4,151,126 (1979)
105. Howard, E. G.: US Pat. 3,480,530 (1977)
106. Schoppel, W., Reichert, K. H.: Macrom. Chem. Rapid Commun. *3*, 483 (1982)
107. Bogomolova, N. M. et al.: Proc. IX Symp. Synthesis and Properties of Graft polymerized on Fillers Polyolefins, Chernogolovka, Moscow District, p. 64 (1982)
108. Grigorian, A. et al.: ibid. p. 90
109. Buniat-Zade, A. A. et al.: ibid. p. 130
110. Howard, E. G. et al.: Ind. Eng. Chem. Prod. Res. Dev. *20*, 429 (1981)
111. Howard, E. G., Glasar, B., Collette, J.: Soc. Plast. Eng. Nat. Techn. Conf. on High Perform. Plastics, USA, Oct. 1976, p. 36
112. Enikolopian, N. S.: J. Nation. Chem. Soc. (USSR) XXIII, No 3, 243 (1978)
113. Gevorgian, M. A. et al.: ref. 107, p. 136
114. Fridman, M. L. et al.: ref. 107, p. 113
115. Fridman, M. L. et al.: Dokl. AN SSSR *225*, 1185 (1980)
116. Fridman, M. L. et al.: Plast. Massy No 2, 17 (1982)
117. Stalnova, I. et al.: ibid. No 3, 15 (1982)

118. J. P. Presentation No. 47–1490, Solvay and Cie.
119. Fridman, M. L.: Doctorate Thesis Moscow, Institute of Chem. Phys., AN SSSR, 1982
120. Bunina, L. et al.: ref. 107, p. 116
121. Sizova, M., Karmilova, L., Volfson, S. A.: Thes. Nat. Sci. Conf. Filled Polym. Mater. Moscow: NIITEKhIM 1982, p. 35
122. Ponomarenko, A. T., Shevchenko, V.: Uspekhi Khimii *70*, No 8, 1336 (1983)
123. Galashina, N. et al.: Thes. VI Intern. Microsymp. Polymer Compos, Budapest, 1983, pp. 9, 60
124. Sichel, E. K. (Ed.): Carbon-Black Polymer Composites, New York: Marcel Decker 1982
125. Yamashita, S., Kohjiya, S.: J. Appl. Polym. Sci. *17*, 2935 (1973)
126. Yoshida, H., Higashida, F.: ibid. *18*, 939 (1974)
127. Atsuta, M., Turner, D. T.: Polym. Compos. *3*, 83 (1982)
128. Europ. Plast. News *7* (3), 13 (1980)
129. ICI Technic. Inform. 1982
130. Wenig, W., Meyer, K.: Colloid. Polym. Sci. *258*, 1009 (1980)
131. Robson, P., Sandilands, G., White, J.: J. Appl. Polym. Sci. *26*, 3515 (1981)
132. Gupta, A. K. et al.: ibid. *27*, 4669 (1982)
133. Alle, N. et al.: Rheol. Acta *20*, 222 (1981)
134. Ford, A., Goettler, K.: ACS Polym. Prepr. *15*, 451 (1974)
135. J. P. Presentation No 49-19097, Sumitomo Chemical Co.
136. J. P. Presentation No. 48-10187, 48-12345, 47-23438, Mitsubishi Petrochemical Co.
137. Mod. Plast. *56* (12), 44 (1979)
138. Bader, M. G.: Reinforced Thermoplastics in Handbook of Composites (Ed. Kelly, A., Mileiko, S. T.) v. IV, New York, Elsevier 1983
139. Vinogradov, G. V.: in: Uspekhi reologii Polymerov (Advances in Polymer Rheology) Moscow: Khimiya Publishers 1970, p. 98
140. Fridman, M. L., Prut, E.: Uspekhi Khimii, accepted for publication (1984)
141. Fridman, M. L.: Thes. XI Nat. Symp. Rheology, Suzdal, USSR Academy of Sciences Publication, v. 1, 194 (1982)
142. Fridman, M. L., Vlasov, V., Malkin, A. Ya.: Mekhanika Polimerov *3*, 535 (1977)
143. Han, Ch.-D.: Rheology in Polymer Fabrication, London: Academic Prers, 1976
144. Popov, V. et al.: Fabrication of Plastics (in Russian), Moscow, NIITEKhIM Publication, No 1, 1981
145. Fridman et al.: Rheology of Polymers and Dispersed Systems, Minsk, Byelorussian Academy of Sciences Publication, v. 1, p. 123 (1975)
146. Christi, M. A., Darlington, M., Smith, G. R.: Proc. Intern. Conf. Compos., England 1978
147. Xavier, S. F., Tyagi, D., Misra, A.: Polym. Compos. *3*, 88 (1982)
148. Jensen, M., Whisson, R. R.: Polymer *14*, 193 (1973)
149. Campbell, D., Quayyum, M. M.: J. Mater. Sci. *12*, 2427 (1977)
150. Kantz, M. R., Newman, H. D., Stigale, F. H.: J. Appl. Polym. Sci. *16*, 1249 (1972)
151. Haskell, W. E., Petrie, S. P., Lewis, R. W.: Polym. Comp. *4*, 47 (1983)
152. Stephenson, R. S., Turner, S., Whale, M.: SPE ANTEC Techn. Papers *23* (1977)
153. Rubin, I. R. Injection Molding Theory and Practice, New York: Wiley, 1972
154. Plast. Techn. *26* (8), 13,102 (1980)
155. Plast. Mod. Elast. *33* (3), 28 (1981)
156. Plast. World *38* (7), 87 (1980)
157. Bryk, M. T., Chubar, T. V., Kardanov, V. K.: Thermal and Thermooxidative Degradation of Filled Polymers, in: Itogi Nauki i Tekhniki *17*, 225 (1982)
158. Malers, J., Kalvin, M.: Modification of Polymer Materials, Riga: Zinatne *3*, 53 (1972)
159. Cullis, C. F., Laver, H. S.: Europ. Polym. J. *14*, 575 (1978)
160. Jellinek, H. H. et al.: J. Polym. Sci. Polym. Chem. Ed. *17*, 1493 (1979)
161. Kovarskaya, L. B., Sandiarovsky, A. T.: Plast. Massy No 8, 37 (1971)
162. Lin, D. et al.: Thes. I Nat. Conf. Compos. Mater., Tashkent, v. 2, 108 (1980)
163. Schneider, H. S., Reichert, W., Thinius, K.: Plaste und Kautschuk *17*, 310 (1970)
164. CIBA-GEIGY Publ. No. 31124/e.
165. Elson, V. G. et al.: Vysokomol. Soed. *B22*, 494 (1980)
166. Logvinenko, P. H., Goronovsky, G. A.: ibid. *A22*, 812 (1980)

167. Simpson, M. B.: Kautsch.Gummi Kunstst. *33*, 83 (1980)
168. Iida, T., Goto, K.: J. Polym. Sci. Polym. Chem. Ed. *15*, 2427 (1977)
169. Iida, T., Goto, K.: Ibid. p. 2434
170. Kozorezov, Ya. I., Sekho, A.: J. Appl. Chem. USSR No 7, 1614 (1983)
171. Piggott, M. R.: J. Mater. Sci. *13*, 1709 (1978)
172. Piggott, M. R.: Load Bearing Fiber Composites, Oxford; Pergamon Press 1980
173. Titow, W., Lanham, B.: Reinforced Thermoplastics, London: Applied Sci. Publ. 1975
174. Folkes, M. J.: Short-Fiber Reinforced Thermoplastics, New York: John Wiley 1982
175. Utracki, L. A., Fisa, B.: Polym. Compos. *3*, 193 (1982)
176. Deryagin, B. V., Krotova, N. A., Smilga, V. P.: Adhesion of Solids, p. 282, New York, Consultants Bureau 1978
177. Plueddemann, E. P., in: Additives for Plastics, Seymour R. B., ed., vol. I, New York; Academic Press 1977, p. 123
178. Monte, S. J., Sugerman, G.: ibid., vol. 2, p. 63
179. Santrach, D.: Polym. Compos. *3*, 239 (1982)
180. Emmett, A. F.: SPE RETEC, Arkon, Sept. 23, 1970
181. Matzeg, R., Osborne, A. D.: SPE ANTEC, Techn. Papers, *27* (1981)
182. Gorelik, V., Berner, V., Volfson, S. A.: Plast. Massi (Rus), accepted for publication (1985)
183. Plast. World *38*, N 7, 90 (1980)
184. Reinf. Plast., *24* N 12, 382 (1980)

Subject Index

Bei Fragen zur Produktsicherheit wenden Sie sich bitte an:
If you have any questions regarding product safety,
please contact:

Walter de Gruyter GmbH
Genthiner Straße 13
10785 Berlin
productsafety@degruyterbrill.com